THE WORLD SUGAR ECONOMY IN WAR AND DEPRESSION, 1914–40

THE WORLD SUGAR ECONOMY IN WAR AND DEPRESSION 1914–40

EDITED BY

BILL ALBERT AND ADRIAN GRAVES

ROUTLEDGE
London and New York

First published in 1988 by
Routledge
11 New Fetter Lane, London EC4P 4EE

Published in the USA by
Routledge
in association with Routledge, Chapman & Hall, Inc.
29 West 35th Street, New York, NY 10001

Printed in Great Britain by Billing & Sons Ltd, Worcester.

British Library Cataloguing in Publication Data

The World Economy in War and Depression, 1914–40
    1. Sugar industries. Sugar trades. Sugar
    industries and trades, 1800–1940
    I. Albert, Bill, *1942-*   II. Graves, Adrian
    338.1′736′09
ISBN 0-415-00127-7

Library of Congress Cataloging-in-Publication Data

ISBN 0-415-00127-7

Contents

Preface

support for the conference from the Economic and Social Research Council, the Nuffield Foundation, British Sugar plc, the Department of Economic and Social History at the University of Edinburgh and the School of Economic and Social Studies at the University of East Anglia is gratefully acknowledged.

Bill Albert and Adrian Graves

List of Contributors

Bill Albert, University of East Anglia, Norwich, U.K.

Shahid Amin, St. Stephen's College, Delhi, India.

Edward Beechert, University of Hawaii, Honolulu, U.S.A.

Peter Boomgaard, Koninklij Institut Voor De Tropen, Amsterdam, The Netherlands.

Gervase Clarence–Smith, School of Oriental and Africa Studies, University of London, U.K.

Horacio Crespo, Universidad Autónoma del Estado de Morelos, Cuernvaca, Mexico.

Frantisek Dudek, Institute of Czechoslovak and World History of the Czechoslovak, Academy of Sciences, Prague.

Adrian Graves, University of Edinburgh, U.K.

Kusha Haraksingh, University of the West Indies, St. Augustine, Trinidad.

David Lincoln, University of Cape Town, South Africa.

Manuel Martín Rodríguez, Universidad de Granada, Spain.

Yoshiko Nagano, Koryo International College, Aichi–ken, Japan.

M.D. North–Coombes, University of Natal, South Africa.

John Perkins, University of New South Wales, Australia.

Brian Pollitt, University of Glasgow, U.K.

Guy Pierre, Centre D'Etudes et de Recherche Economiques et D'Histoire Economique, Port–au–Prince, Haiti.

Tamás Szmrecsáyni, Universidade Estadual de Campinas, São Paulo, Brazil.

List of Tables

Page

List of Tables

List of Graphs

1

Introduction

Bill Albert and Adrian Graves

At no time in its long and blood soaked history did the commercial production of sugar experience such a thoroughgoing transformation as during the nineteenth century. Of course, this was the century of industrial capitalist revolution when the entire world was transformed, and few commodities or manufacturing processes were left untouched. But, it can be argued that of the tropical products, none was so fundamentally altered as sugar.[1] This change was not only in terms of production methods and structures, labour relations, ownership, and location, but also in the very nature of the product itself and the extent and patterns of its consumption. In order to understand the history of sugar in the two decades following the Great War it is essential to begin with a brief account of the structure of the international sugar economy on the eve of that conflict and to consider some of the principal forces which had shaped that structure.

The nineteenth century

Production

In 1839 it is roughly estimated that total world production of sugar was 820,000 tons, 95 per cent of which came from cane growing countries.[2] By the late 1880s output had grown to over 5.5 million tons. The most spectacular element in this expansion had been the growth in beet sugar, which in 1889 accounted for more than 60 per cent of world production. Subsequently cane recovered some of the lost ground and by 1913, when 18.2 million tons of sugar were produced, the proportion contributed by cane and beet was about equal (Graph 1.3). In many respects this competition between temperate and tropical sugars was a key feature in the commodity's development. The expansion of heavily protected beet sugar

1

put great pressure on cane growers to modernize in order to lower their costs, for especially from the mid–1880s prices fell sharply. At the same time the technical and scientific advances associated with the development of the beet sugar industry in Europe and industrial development here generally gave cane growers the tools with which to compete. The fact that the rate of growth of cane output increased after 1885 suggests that many cane producers availed themselves of the new technology, although the timing and relative success of technical change varied considerably according to a complex of local circumstances, including ecological conditions, existing structures of landholding, government policy and access to markets, capital, and labour. While national and regional differences were important, nonetheless, it can be argued that by WWI, in both Europe and many cane growing areas milling and fabrication had become modern and in most cases large scale industrial processes.

This technical change tended to impose other related changes on sugar growers. One of the most important was on the size of enterprises. To operate the new, more efficient machinery economically demanded a larger scale of production and this plus the fact that many growers could not easily raise the substantial amounts of capital needed to finance the new ventures led to widespread consolidations and a substantial increase in direct foreign control. By 1913 while output in most cane growing countries had increased the number of mills had fallen. In some places, such as Cuba, this was accompanied by the development of central milling, with cane grown by independent or semi–independent farmers. In Hawaii and Peru, both dependent on irrigated land, it was the large scale integrated agro–industrial estate which came to dominate. With an increase in size there was also a move toward corporate forms of organization, often associated with the growing dominance of foreign capital.

The other major change for cane sugar producers in the nineteenth century was in the relations of production. Until the early 1800s sugar production in the New World had been invariably associated with black chattel slavery. This began to change with the successful slave revolt in the most important sugar producing French colony of St. Dominque (Haiti) in the 1790s. The slave trade was abolished by most countries a few years later and this was followed by emancipation in the British colonies in the late 1830s. Although similar measures were taken by other colonial powers, it was a fairly slow process and it was not until 1888/89 that slavery was finally ended in Cuba and Brazil.

The slave was not, however, immediately replaced by the wage labourer. Planters may have lost the right to hold men and women as property, but they did not lose their slave–holding attitudes about how to deal with their workers. In short, plantation work remained, to say the least, extremely unattractive. Furthermore, the larger new plants demanded more cane and, therefore, more workers. As many planters had low expectations about their ability to draw upon or create a market for labour, they frequently resorted to bringing in indentured workers from Asia or the Pacific Islands. These semi–servile workers provided a convenient bridge between slavery and free wage labour, the latter which was only really beginning to become common by the end of the century.

It is clear that although planters were faced with need to adapt their production methods to an international economy increasingly dominated by the logic of industrial capitalism, they were extremely reluctant to accept a key element of that logic – free wage labour. As noted, this was to some extent a product of their slaveholding past, but it also had to do with the essentially uneven character of technical change as between factory and field. Moreno Fraginals, writing on Cuba, observes[3], 'Through most of the past century sugarmen concentrated on the manufacturing aspects of production and much less on the agricultural. All the world's sugar producers did likewise.' This was because the new processing and milling machinery, derived in the main from the refining and beet industries of Europe, had to be adopted if cane producers wanted to remain competitive in the world market. In the cane fields, however, the extreme diversity of local agricultural conditions made the transfer of technology more problematical. There were some important innovations, such as movable rail lines and the steam plough, but these were borrowed from a more generalized pool of technological advance and did not result in a transformation of the labour process in cane production. This technological divide gave cane sugar production its dual structure which was apparent virtually everywhere by 1914 – a highly scientific and technologically advanced manufacturing sector wedded to an agricultural sector, which despite improvements, still relied on gang–worked labour which had changed hardly at all for three hundred years.

An early nineteenth century observer transported to a sugar estate in 1900 would have recognized little that he saw. The mill and the factory were completely changed into large scale industrial plants and the relatively uniform, centrifugal sugar they produced bore no resemblance to the loaf and many other varied qualities of sugars to which he would have been accustomed. Only when he left the the heat and noise of the mill and wandered out into the fields would he have seen a familiar sight, because the very heart of the process, the harvest, continued to depend on men with machetes.[4]

Trade

Sugar, a major trade good in the mercantilist age, had always been the most political of products. Little changed in the so–called era of Free Trade. Initially the production of sugar from beet was only possible if the lower cost cane sugar was excluded, and throughout the nineteenth century

European producers relied on the state to protect their industries. As competition increased tariffs alone proved to be inadequate and a complex system of export bounties was established in many countries.[5] These were not finally suppressed until the Brussels Convention of 1902. While this made conditions for cane producers somewhat easier, in fact the only major free market in this period remained Great Britain, which accounted in 1913 for about a quarter of the world's and three–quarters of European sugar imports,[6] of which most of the latter (75 per cent) came from Germany and Austria–Hungary. The United States was the other principal sugar importer (25 per cent of total imports) and here by WWI the offshore possessions of Puerto Rico, Hawaii and the Philippines had free access and Cuba special tariff concessions. In the other important area of trade, Asia, it was the Indian market and to a lesser extent those of China and Japan which were of prime significance, with Java by far the most dominant exporter (73 per cent of Asian sugar exports).

A simple statistical view showing Cuba (30 per cent), Java (15 per cent), Austria–Hungary (12.7 per cent) and Germany (12 per cent) as the world's leading sugar exporters in 1913 fails to give a clear picture of the full importance of the production of this commodity. Many countries with much smaller industries relied as heavily on sugar as did such large producers as Cuba or Java. In Hawaii, Puerto Rico, and Trinidad, as well as many other Caribbean islands, in British Guiana, Mauritius, and Fiji sugar was the principal export. Furthermore, there were areas of other countries in which sugar was the dominant industry, for example, in the Mexican state of Morelos, in Pernambuco in northeastern Brazil, the island of Negros in the Philippines, the north coast of Peru, the state of Queensland, the northwest of Argentina, and on the coast of Natal.

Although cane sugar was being produced in independent countries, it also continued to form an important part of the matrix of nineteenth century colonial economic systems, both formal and informal. As such it did little to foster more broadly based progressive capitalist transformations. The sugar regions became increasingly dependent on this single crop and at the same time equally dependent on the import of manufactured goods and in some cases food as well. This situation was exacerbated by the fact that from the mid–1880s the terms of trade between sugar and manufactured goods showed a strong downward trend.[7] Having, moreover, to rely on imported technology, an often capricious, if not manipulated foreign market,[8] and employing large numbers of poorly paid semi–skilled workers, the commercial production of sugar for all its apparent promise as a money earning export became in most cases a developmental dead end. These problems were compounded in those cases where direct foreign control led to profits being accumulated abroad.

However, while the long–run gains for producing regions are in doubt, the consumers, especially in Europe and the United States did benefit from the massive growth in world sugar output. During the nineteenth century the price of sugar fell, with two major breaks in the late 1840s and mid–1880s. There were distinct national differences depending on exchange rate movements, real wages, and the extent of taxation on sugar, but it would also appear that in real terms the price of sugar declined substantially. In Britain as early as 1850 sugar had ceased to be a luxury and had become an important item in the working class diet.[9] As prices fell so consumption rose, reaching a peak of about 42 kilos per person in 1912.[10] Similarly high levels were found in Denmark and the United States, whereas in Germany and France, although consumption did increase, whether for reasons of cost or culture, people ate only half this amount. Nonetheless, by WWI, sugar had become a major staple food in most developed countries. In Britain, for example, it made up about 17 per cent of calorie intake for the entire population, probably being substantially higher for the working class.[11] Sidney Mintz has observed, '... the ever–rising consumption of sugar was an artifact of intraclass struggles for profit – struggles that eventuated in a world–market solution for drug foods, as industrial capitalism cut is protectionist losses and expanded a mass market to satisfy proletarian consumers once regarded as sinful and indolent.'[12]

WWI and the interwar years

Trade

The key importance which sugar had assumed for the British is highlighted by the fact that it was the first commodity which the government moved to protect when war broke out in August 1914.[13] The Royal Commission on the Sugar Supply was established on the 20th of that first month of the conflict and for the remainder of the war this most important free market was suspended with shipping, prices, purchasing and distribution all being firmly controlled.[14] Although prices continued to rise it was claimed that, 'Unquestionably the Commission by its forward purchases in a rising market, made possible a retail price in the United Kingdom, well below the level that would have been determined by private enterprise.'[15] A further stricture on the international market came after the United States entered the war in 1917 and joined with the British to form the International Sugar Commission. It was felt that '... some form of cooperation with the Allies was imperative so that there would be no further competition among them in securing sugar supplies. The soaring prices which must of necessity rule if the law of supply and demand were allowed full play would thus be eliminated.'[16] In other words, sugar producers were not allowed to reap the full profits which might have accrued to them because the principal buyers felt it was not in their interests to allow it. While it may be argued that this was simply a product of wartime conditions, which, of course, it was, at the same time it served as a graphic demonstration of the unequal relationship between tropical sugar producers and their metropolitan customers.

Despite Allied controls, the price of sugar did rise during the war, and many sugar growers made substantial profits. If, however, the trend in the Latin American figures are indicative,[17] the terms of trade moved strongly against primary producers during these years, suggesting that profits from sugar, already curtailed by a buyer's cartel, would not have been enough to recompense the respective economies for higher priced imports.

Both total world sugar production and exports fell significantly (about 16 per cent each) during the war years, entirely because of the sharp decline of European beet sugar. Output of beet which had reached a pre-war high of over 9 million tons in 1913/14 was down to but 4.4 million tons five years later, while cane sugar, which had increased by an average of 4.6 per cent per year, from 7.7 million tons to 9.6 million tons, failed to make up the shortfall. It is often claimed that the wartime conditions stimulated a massive expansion in cane production, but this was not apparent by the armistice. Only some of the cane growing countries were able to increase output in the period (1913/14–1918/1919) and on a world level it was the incredible growth in Cuban production of 1.46 million tons (9.1 per cent per year) which accounted for the vast majority (76 per cent) of the rise in cane sugar output. Secondly, despite the three fold rise in prices, the rate of increase in cane production during the war was exactly the same as it had been between 1900 and 1913, and considerably less than in the five years before the war (6.3 per cent per year). The reasons why more substantial gains were not made is unclear, although it may have had something to do with Allied restrictions on shipping and the export of machinery.

The war ended in 1918, but it was not until December of the following year that the the Anglo–American control of the sugar market was lifted, with the disbanding of the US Equalization Board.[18] This together with fears of a bad Cuban harvest (in fact in 1919/20 Cuban production fell by about 7 per cent and total world output by 12 per cent) fuelled speculative fever in what was in any case a booming market for most primary commodities. Prices reached a record peak of 23 cents/lb (137s 9d/cwt) in May.[19] This was more than three times higher than the average price in the previous year. In Cuba this sparked off the so–called 'Dance of the Millions' which saw sugar–based speculation engulf the entire economy, with disastrous results when the bubble burst.[20] The response on the part of the sugar producers was understandable. Already encouraged by wartime shortages they were now given an added incentive to plant more cane. But almost before they could get it into the ground prices began to fall, hitting 4.375 cents/lb (27s 6d per cwt) by December. The slide was to continue until late in 1922, as beet sugar began to make a recovery and total sugar production rose by more than 2 million metric tons between 1919/20 and 1920/21.

But although the fall from the heights of May 1920 created serious problems from those who had borrowed money at that time, for the world's sugar producers as a whole recovery came relatively quickly and prices remained fairly buoyant until 1925, despite a steep rise in the production of both cane (by over 7 per cent per year between 1919/20 and 1925/26) and beet sugar (12 per cent per year). Consumption too had risen over this period[21], but by 1925 it was being outstripped by the increase in output and there was substantial price fall of about 40 per cent in that year. Whether because of falling costs, protected markets or simply a rosy view of future prospects, the collapse in prices did not bring the increase in world sugar production to a halt, although the rate of growth did slow considerably in the next four years, to 3.6 per cent for cane and only 1.8 per cent for beet. With prices falling, world sugar stocks going up by about 12 per cent a year from 1925[22] and no sign that supply and demand would be harmonized it was argued by many observers that the international sugar economy was entering a period of severe crisis from about the mid–1920s. This can be seen as part of a more generalized crisis which affected many primary products from about this time.[23] For example, Timoshenko's agricultural indices (1923–5 = 100), indicate a 70 per cent fall in prices from 1925 to 1929 and a 75 per cent increase in stocks.[24]

But this crisis and the much more difficult problems experienced in the 1930s did not affect all sugar producers in the same way. This had to do in part with the way in which they responded to changing conditions, but it was also a function of their access to specific markets, for although we may speak of a 'world market' for sugar, there were in fact multiple markets. For example, by the late 1920s it was estimated that only 25 per cent of sugar was being sold in non–preferential or non–protected markets.[25] In other words, only a quarter of the sugar produced was traded on the free market. The proportion had been reduced to only 10 per cent by the end of the 1940s.[26] This disarticulation of the international market was in turn largely responsible for the failure of the price mechanism to operate so as to reduce supply. Clearly apparent in the prewar years, this became a far more serious and widespread problem after 1918, when the world economy itself failed to regain the relative balance and cohesion that it had achieved in the nineteenth century, as '... the movement towards protectionism became more marked, [and] international considerations were increasingly subordinated to national monetary and employment policies made necessary by post–war reconstruction and, later, by the onset of a world depression.'[27]

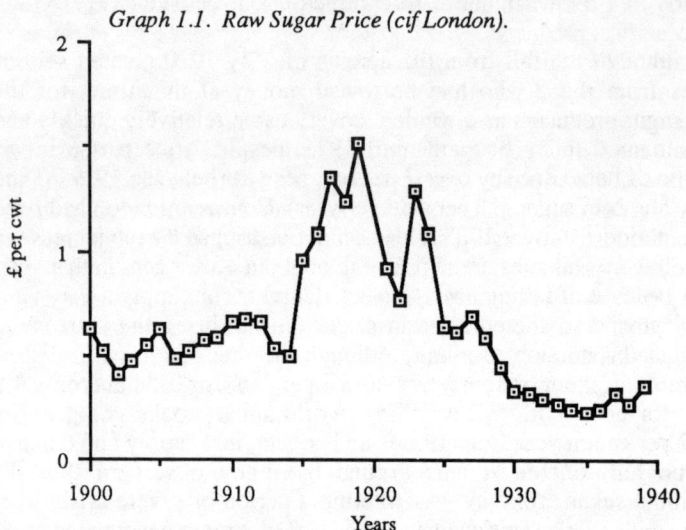

Graph 1.1. Raw Sugar Price (cif London).

Source: U.N., F.A.O., *The World Sugar Economy in Figures, 1880–1959*, (Rome, 1960).

For many sugar producers it was the commercial policies of the two main importers which were of key importance, and in both cases the trend was toward greater restriction. Perhaps potentially the most serious change was that effected by Britain for this had been the one major open market before 1914. After the war her own economic problems and the weakening of the international economy pushed Britain increasingly to modify and eventually to abandon her free trade position. With respect to sugar this took the form of colonial preference and official support for the creation of a domestic beet sugar industry. The latter move had much to do with the desire to increase the country's self-sufficiency after the experience of WWI, when it became clear how dependent Britain was on Europe.[28] Initially the wartime tariff and an excise tax rebate was used to protect the new beet industry, but in 1924 a more formal system of subsidies was adopted, as the country followed the European model. This was augmented by the Sugar Industries Act of 1936 and the formation in that year of the British Sugar Corporation.[29] Production of beet increased from virtually nothing before the war to about 450,000 metric tons by 1930/31, with an obvious effect on the country's demand for imports.

The other limitation on imports into Britain was the continuation of wartime tariffs and the granting of preference to imperial cane producers from 1919. The conditions were modified from that time, most significantly in 1928 when duties on refined sugar imports were raised [30] and European beet producers, whose exports were almost always of refined sugar, were thereby cut out of the market. By the 1930s almost all of the country's imports were coming from cane growing regions. Moreover the share of empire producers rose from 3.7 per cent in 1913 to 28.3 per cent in 1930 to over 50 per cent by the end of the decade.[31] It is curious that although British sugar policy was restrictive and led to the creation in the late 1930s of a domestic producers cartel (in an agreement between refiners and the beet industry) overall it seems to have worked to the advantage of many cane sugar producers, firstly because a great number were within the British Empire, and secondly because the heavily subsidized beet industries, which had completely dominated the British market were now eliminated as competitors. However, simply having access to this market did not insulate exporters from the devastating impact of lower world prices.

The other major world importer was the United States, and here protection had always been a major element of economic policy. The tariff was raised to very high levels (by about 60 per cent) in 1921 and again a year later on the urgings of domestic growers. But it was not mainland producers who benefited, for with prices falling they found it difficult to compete with the lower cost industries in the duty–free off–shore dependencies of Hawaii, Puerto Rico and the Philippines, which increased their deliveries to the US market by over 80 per cent (from 1.04 million to 1.88 million tons) between 1919 and 1927.[32] In 1930, the Smoot–Hawley Tariff raised the duty to 2 cents per pound and despite the fact that this included the continued 20 per cent preference for Cuba, it led to a major crisis for the industry here, so dependent was it on the US market. Sales fell from 4.1 million tons in 1929 to only 1.6 million in 1933.[33] Not only were problems created for the entire Cuban economy, but 'to the extent that Cuban sugar was not saleable in the United States, exports were pushed more aggressively elsewhere, with disastrous effect on the 'world' market price'[34]

But problems facing US sugar producers continued to become more acute, and in 1933 the US Tariff Commission concluded, that Cuba simply lowered the price to the level needed to get into the US market and that this was hurting both Cuban and American growers.[35] There was also concern about the impact on US economic interests of the severe Cuban crisis, which had led to revolution in 1933. The Jones–Costigan Act of 1934 changed the basis for control of imports to a quota system and under one of its provisions the President lowered the tariff on Cuban sugar by half a cent to 1.5 cents. Later in that year under a reciprocal trade agreement the tariff was further cut to only 0.9 cents.[36] Although all was far from perfect, the new legislation significantly reduced the difficulties facing the Cuban industry and in doing so undoubtedly also eased pressure on the international market. It must be remembered, however, that although both US and British sugar policies had a general impact on the world market, their specific effect on individual sugar producing countries varied considerably and can only properly be studied within the context of each industry's development.

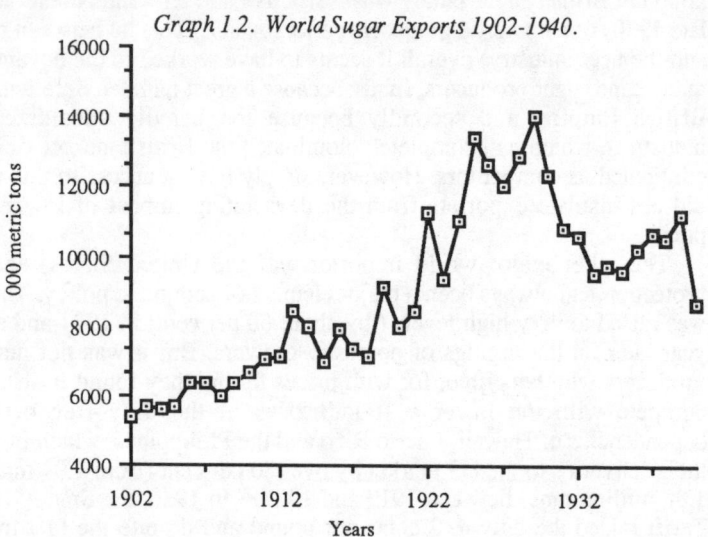

Graph 1.2. World Sugar Exports 1902-1940.

Source: U.N., F.A.O., *The World Sugar Economy in Figures, 1880-1959*, (Rome, 1960).

Besides the various national sugar policies, during the interwar period there were also a number of attempts to get an international agreement to reduce production and export. The Cubans made efforts to arrive at some accord on cutting back output in 1927 with the industries in Germany, Poland and Czechoslovakia. But, because of the failure to involve other producers, especially Java, this came to nothing, as for the same reasons did another attempt in 1929.[37] With the general crisis biting harder in the 1930s, there was more widespread interest in an accord and in 1931 the Chadbourne Agreement was signed by seven major producers, who in 1929-30 accounted for about 50 per cent of world production.[38] However, although the signatories did reduce plantings, like its predecessors this agreement had little success. This was due to the failure to involve the major importers and the encouragement given to sugar production by some of the non–participating countries. By 1933–34 the adherents to Chadbourne were producing only 25 per cent of world output. A potentially more successful international convention was signed in 1937 by twenty–one nations, including Britain and the US. The coming of war meant that the system of export quotas proposed was never fully tested, and it would seem that the agreement had little or no effect on the world market.

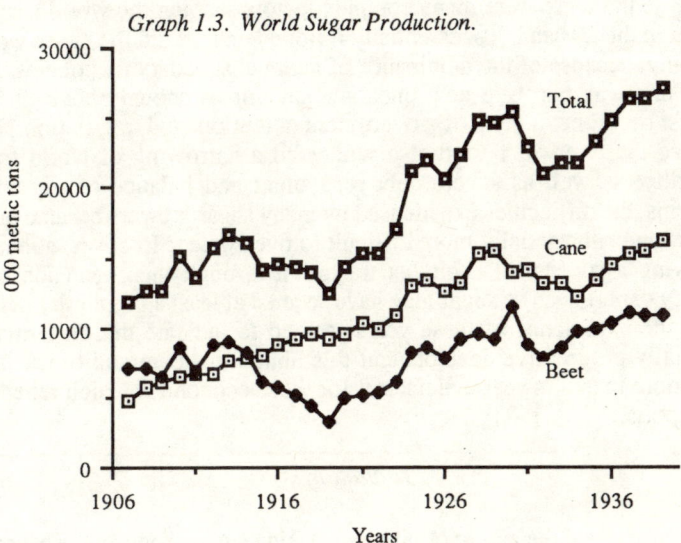

Graph 1.3. World Sugar Production.

Source: U.N., F.A.O., *The World Sugar Economy in Figures, 1880–1959*, (Rome, 1960).

As argued above, the crisis for most sugar industries began after 1925. In the years from 1924 to 1929 the average price decline had been about 20 per cent a year. Conditions became significantly worse in 1930 when the there was a further 27 per cent fall. Subsequently the price continued to drift down until 1936. But, it was not until three years later with the outbreak of war that a sustained improvement came. On a world scale production began to be cut back from its peak in 1930/31, with major reductions seen in Java, Cuba, and Czechoslovakia, as well as some of the other Chadbourne countries. Sugar stocks also were significantly depleted. The move was, however, shortlived, for despite there being no substantial improvement in the world price, both beet and cane production began to rise once again from 1935. This was fairly evenly spread across regions, although Cuba, Germany, and India all made large contributions to the overall increase. The reasons why supply response once again seems to have been so perverse lies as before in the fact that the world price did not reflect the price facing growers either in their own protected markets, as was the case in Germany and India, or because they had preferential access to certain markets, as Cuba did. In short, the general crisis facing the world sugar economy could not be resolved because many governments sought, not without reason, to shield their producers from the full effects of the crisis.

While the there were many complex factors shaping the world's sugar market in the 20s and 30s, essentially it floundered, as did the entire world economy, because of the dominance of nationalist economic policies. For sugar this was hardly a new phenomenon, for as shown above, it had always been the object of government attention and regulation. But because the interwar period also witnessed a narrowing of world trade generally, as well as severe debt repayment and balance of payments problems, the difficulties experienced by many sugar growers became more serious and substantially more difficult to overcome. Moreover, although achieving a successful capitalist transition through the expansion of a primary export such as sugar may have seemed at least a possibility before 1914, the experience of these years seemed to indicate that in terms of materially progressive development this ubiquitous commodity was still little more than a 'sweet malefactor' for those economies which relied on her exports.

Production [39]

Having surveyed the nature of problems facing sugar producers because of changing market conditions in the interwar period, we now turn to examine the responses of various industries to the crisis. A major difficulty with such a survey is that throughout the world sugar production was carried out under radically different conditions, in countries with markedly dissimilar histories, socio–economic and political structures and positions within the world economy. This makes it extremely difficult to generalize about the strategies of the industry as a whole, although it is possible to draw useful contrasts and comparisons. We shall consider a number of key variables, including changes to productive structures, technological innovation, access to capital and credit, the problem of labour, whether wage–labour or peasants, and finally, the role of the state.

Structural changes

The crisis of the interwar period tended to accelerate the process of restructuring in both beet and cane sugar which had become so evident in the decades before WWI. Generally this meant an increase in both the concentration of ownership and the centralization of production and distribution. There was, however, an enormous degree of variability in both the extent of capital concentration and the forms that it took. In Hawaii, for example, the industry became almost completely vertically integrated under the control of six companies, whereas, at the other end of the spectrum, Cuba actually saw the weakening of its largest sugar companies in this period and the ascendancy of small producers. For many countries, such as Peru, Mexico and Guiana, the concentration of mill ownership in this period was a continuation of a process which had begun in the last great sugar crisis of the 1880s. In South Africa and Australia and much of the West Indies, there were few appreciable changes in the structure of ownership, but there was growing centralization of production. Finally, there was the dynamic emergence of essentially new industries in Britain and India, which in both cases were highly centralized and concentrated.

In many countries, the process of concentration was confined to the industrial side of production. In Germany three large combines of sugar millers controlled 90 per cent of the country's sugar output by the late 1920s and a similar process was evident in Czechoslovakia and in cane producers such as Mexico or Argentina. Almost invariably (Peru and Hawaii being major exceptions), concentration in mill ownership was accompanied by either the break up of large cultivation units and/or the entry of many small cane growers. This was consistent with the centralization of milling processes, which to achieve the needed economies of scale required a larger cane supply. Relying on outgrowers also helped shift a large part of the risk from millers to farmers, a particularly advantageous strategy for the latter group in these trying times. This also partly explains the increasing incorporation of peasants into cane and beet production in some regions. In response to this there was a tendency for cane farmers to organize to counter the oligopsony power of the millers. As with production so the marketing of sugar was also taken over by fewer, larger companies. In London, one of the two major sugar markets of the world, the number of sugar brokers fell from seven to one in the interwar period. In many countries, sugar producers or the state formed marketing bodies or authorities to distribute their output, both on domestic and foreign markets. All these moves toward greater concentration clearly reflected and at the same time contributed to the narrowing of the international market for sugar.

It is important to stress that while most of the sugar producers were responding to world market conditions, local factors, including the prior history of sugar production in the regions, played a determining role in shaping the industry. In Germany for example, the rapid concentration of mill ownership in the interwar period, was attributable as much to major losses of export markets and territory in the east following WWI. This change was also greatly influenced by the fact that in a period of high inflation mill owners could get access to cheap credit in their capacity as owners of refineries. At the same time, the pace and form of structural change could be strongly conditioned by landholding systems, as it was in northeastern Brazil or the Philippines. But possibly the single most important factor which acted upon the course of change was national political exigencies. As will be shown below, in almost every producing country it was the state which set out many of the key economic parameters within which their sugar industries had to operate.

Changes in technology

The need to reduce costs and raise profitability gave rise to broadly based changes in the technology of sugar production at all levels in the interwar period, although many innovations were merely developments of those originating in the late nineteenth century. It is extremely difficult to generalize, but it is certainly true that the manufacturing processes of sugar boiling, panning and refining advanced considerably. In the mills capacity was not only expanded enormously, but crushing technology appears to have reached its height in this period with mills capable of 98 per cent rates of extraction. One interesting consequence of such improvements was the transfer in technology to which it gave rise. When progressive sugar industries re–equipped with the most advanced technology, they sometimes exported their now redundant machinery to less advanced producers. This process was also often an intra–industry phenomenon. Moreover, economies which experienced a sharp contraction due to the restructuring of the international sugar economy off–loaded their redundant equipment onto the developing sugar regions. The rapid expansion of India's sugar industry, for example, was serviced by the transfer of such equipment from Java, Mauritius and elsewhere.

But it was on the agricultural side, particularly for the cane producers, that the most dramatic changes occurred, redressing to some extent the technological imbalance between field and factory so apparent before WWI. The growing reliability and robustness of petrol driven engines allowed for the widespread introduction of tractors and other vehicles into field operations, along with gang disc ploughs, auto rotary hoes, planting machines, interspace cultivators and weeders and ratooning implements. Other major innovations included the introduction of track laying vehicles to improve the speed and safety of transporting cane to mills or in less developed economies, the significant transition from wooden wheels to rubber tyres. The first promising invention of a cane harvester in Australia, the famous Falkiner Harvester, occurred at this time, and more

14

generally attention was given to improving the irrigation, drainage, and fertilization of sugar crops.

While all the aforementioned changes were important, in terms of increasing cane yields the single most important change was probably the development of high yielding cane varieties, especially POJ 2878. This variety, developed in Java, spread to many other regions in the 1930s and led to spectacular improvements in land productivity. For example, in Java itself the increased cane yield for one year with the new variety was enough to repay the money spent by the Java Sugar Experiment Station in forty years.[40] Not surprisingly, higher yields gave a further twist to the problems of world oversupply. In this, as with state intervention, 'solutions' could easily become part of the problem they were devised to address. Cane growers also benefited from advances in the fields of soil science, entomology and cane pathology. The emergence of government experiment stations in many countries during the nineteenth century was consolidated and accompanied by a flourishing literature which was disseminated through international journals devoted exclusively to the problems of sugar production. A wider forum for this work was provided by the establishment of the International Society of Sugar Cane Technologists. It is interesting that in this troubled industry there should be such a seemingly high degree of international technological cooperation. It may have been because in a real sense sugar producers were not directly competing against one another but primarily against their own ability to profit by reducing costs to meet a price over whose movement they had absolutely no power. Furthermore, in a world increasingly characterized by politically determined markets and prices there was little to be gained from technological exclusivity. Even when producers did compete directly, as did the Peruvians for their domestic market from the late 1920s, there was never any attempt to curtail the free interchange of technical information.

Though it is evident that it had a major influence on the economics of the industry, it is extremely difficult to calculate precisely the impact of innovations on the international sugar industry in the interwar period. What is clear is that the introduction of new technology was both highly variable between the sugar producing nations, and that there were even significant differentials within regions of the same nation. In some places, such as in Germany, sugar technology actually stagnated over the period. In certain third world regions such as the Philippines, parts of Africa, Latin America and the Caribbean, in general sugar technology advanced but little and remained behind that found in the more advanced economies of Australia and the US. In newly emerging sugar producing regions in southern Africa and southeastern Brazil producers tended to operate on a more sophisticated level of factory and field technology than the older, more established areas. Much of the differential technical response had to

15

do with the availability of capital, and it is to this subject which we now turn.

Finance and Credit

If it is possible to generalize at all on this complex subject, the nature and origins of capital for the sugar industry appears to have depended mostly on whether it was directed into the manufacturing or agricultural sectors and on whether it was in metropolis or on the periphery. Let us turn first to the industrial side of sugar production.

The capital requirements of modernizing factories and mills were enormous. In some countries, the stock exchange was a significant source of capital for this sort of investment. The ascendancy of the miller–cum–planter in milling and refining represented also the rise of the limited liability company in such places as South Africa. It should be stressed, however, that in many sugar producing countries listing on the stock exchange often was a blind for family ownership, such as in the Philippines and in much of Mexico or South America. In others, the existence of the limited liability company reflects a history of bank takeovers and floatations in periods of economic crisis and a broader spectrum of private and institutional shareholders were involved. Nevertheless, private companies did not always resort to the stock exchange to raise capital.

The Colonial Sugar Refining Company in Australia, for example, the country's monopoly refiner and also an important mill owner, financed its expansion and innovation from ploughed back profits. The Spanish sugar industry appears partly to have financed its modernization in the interwar period out of the profits it reaped during WWI. The rapid development of the British industry in the interwar period was also financed largely out of private capital, particularly by Tate and Lyle, from the accumulated profits of its substantial involvement in cane sugar production. Although the financing of the Hawaiian industry was complex, private capital played a dominating role. It should be stressed however, that all of these industries, and by extension their investors, were very heavily subsidized. The British industry received a subsidy of between 140–150 per cent of the world free price of sugar. Subsidies amounting to 1¢ to 2¢ per lb benefited the Hawaiian industry.

Beyond the stock exchange and organic investment, large commercial banks financed capital investment on the industrial side. These banks appear to have been involved at all levels of production in most of the producing countries, but with widely varying degrees of intensity. In many countries the state was crucial in supporting millers and refiners, with state agricultural banks or the central banks and state run industry organizations coming to the rescue of sugar producers, often under preferential conditions, when other sources of capital were not readily available.

Commercial and state banks were not the only sources of credit for the industrial side of sugar production. The Barbados Mutual Insurance Company, for example, was heavily involved in the local sugar industry. In Java industry credit cooperatives founded in the late nineteenth century were an important source of capital for local refining and milling. In India and Mauritius merchant and rentier capital serviced the milling side, including the extension of credits by sugar brokers and agents to the industry. It was also very common internationally for the suppliers of machinery to extend credits to millers and refiners, though this form of financing seems to have diminished in importance as the period advanced, especially after the onset of the Great Depression.

While the industrial side of sugar production required particularly large amounts of capital, the implementation of innovations in cultivation and harvesting required finance as well. In contrast with the financing of the manufacturing side of the industry, small scale farmers relied heavily upon short–term credit to meet the basic requirements of production. Commercial banks and state banks, including a growing number of agricultural banks and state funded rural agencies, were a major source of farm finance in this period, especially in the most dynamic of the sugar economies. In Australia, for example, banks preferred to deal with farmers rather than millers, because it was felt by the commercial banks that the high level of state regulation imposed on millers limited their ability to service debt. The widespread growth of farmers' cooperatives in many countries following the centralization of milling processes, gave rise to cooperative credit societies to service farmers needs. Millers and sugar agents were common sources of farm credit, and in developing countries in particular, this was often associated with high levels of debt amounting to virtual entrapment especially in areas dominated by peasant production, where local landlords and shop keepers also advanced credit.

The variability and differential nature of credit and finance in the international sugar economy in this period was also affected by broader macro economic factors and by the location of sugar industries. As a general rule, the onset of the Great Depression saw a sharp decrease of funds available for sugar production from commercial sources of finance. This proved to be a particular problem for those countries on the periphery who had become dependent upon metropolitan capital. Cuba, probably saw the most dramatic flight of capital following the onset of the Great Depression, but French, Dutch, Portuguese, and British sugar colonies also experienced sharply reduced financial support from metropolitan sources. The effects of the flight of metropolitan capital from the periphery was, with the exception of the British colonies who came under the umbrella of the Empire Preference Scheme, exacerbated to some extent by the realignment of long standing trading patterns in the international sugar economy in the wake of the Chadbourne Agreement in 1931 and the International Sugar Agreement of 1937.

Labour

The availability of credit and long–term finance was obviously of crucial importance for it had a direct affect on the ability of particular industries to introduce new technology and modernize their structures of production. Such changes were prompted by the urgent need to cut costs, and as by far the most significant component of these costs was labour, capital formation and technical improvement therefore meant, as they always had, that the nature of the labour process had to be altered. But such alteration was not an easy or straightforward matter. It involved changes in the recruitment of labour, the structure of the workforce, patterns of work and social conditions. In an industry not known anywhere in the world for the equitable treatment of its workers and where in many regions attitudes towards labour were changing only slowly from a relatively recent slave–holding past, introducing such fundamental changes was bound to result in hardship for workers as attempts were made to shift a large part of the burden of the adjustment to the crisis onto their backs.

As might be expected, there was a considerable diversity in the labour systems found in the various sugar industries, and even within the same industry different regimes were employed. For example, in most countries permanent wage labour, both skilled and unskilled was employed in the factory and mill, while field work was carried out under various systems ranging from independent peasant production and sharecropping to wage labour, whether semi–permanent or migratory. One of the most significant changes in this period was the abolition of indentured labour. The ending of this system was closely associated with the development of central milling, the concomitant reduction in the number of large integrated plantations and the greater reliance on buying in of cane from outgrowers. This in turn led to a dependence on seasonal wage labour, possibly one of the most universal characteristics of the sugar industry whether beet or cane.

Inevitably, in an industry almost universally characterized by a high level of ethnic stratification, a close relationship was maintained between occupational and ethnic differentiation. A reserve army of immigrant, often temporary migrant, workers continued to be a mainstay of the industry in many parts of the Caribbean and Latin America. In other areas, notably, Germany, Australia and Hawaii, high levels of local unemployment led to reductions in the level of immigrant workers in favour of the employment of nationals. With the greater reliance upon family labour in farm production, women and children came to constitute a significant section of the labour force. But they were also important in those sugar economies which had to compete intensively for their seasonal labour supplies. In South Africa, for example, gangs of women cane cutters were prominent in the Zululand industry, and young boys in the 1920s constituted almost 20 per cent of the labour force in that region. Women and children were also very important in the Philippines , according to a very strict division of labour with men, especially in field work duties. Paradoxically, women workers were systematically forced out of the Jamaican industry in this period. Whereas in 1910, they made up as much as 50 per cent of the labour force, by 1940, their numbers had been reduced to barely 20 per cent.

While the requirements of rapidly changing technology produced changes in the structure and composition of the industry's labour force, workers were also called upon, as they always are, to shoulder the cost of lower prices. Almost universally, the onset of the Great Depression saw sugar workers' money and real wages being cut either directly or indirectly. In the latter category were attempts to increase the extraction of absolute surplus labour by the greater use of piece work payments and the butty gang principle (in which returns were directly dependent on the amount of cane cut), the mainstay of harvesting in some countries, particularly in Australia. Contracting industries, such as those of Java and Cuba, experienced major job losses and the successful introduction of new technologies also reduced demand for labour in some areas which, of course, was generally the intention of such innovation.

In many countries, working conditions deteriorated markedly as a result of the efforts of employers to reduce their outlays on social overheads. This was associated with the increasing morbidity of workers and a greater concern on their part about occupational diseases including agricultural workers' phthisis and Weil's disease. While labour legislation was widely enacted to regulate working conditions and pay it often had the effect of imposing control over the labour market to the particular advantage of owners. Neither was the state averse to the use of repression to maintain control over recalcitrant workers or their organizations. In some respects, the hand of both the employers and the state was strengthened in their relations with workers because of the generally adverse economic situation and the many impediments to effective worker resistance.

The universally high level of stratification in the industry, by occupation, ethnicity and gender tended to create or exacerbate divisions and tensions in the workforce so undermining collective action. The increasing seasonality of sugar employment, and the growing trend towards the organization of work around contract gangs also reduced the conditions necessary for collective action. Resistance was also made more difficult because of the threat of unemployment, especially acute during the Great

Depression. Sugar estates were also often geographically isolated from one another, and this could pose serious problems for wider collective action. Moreover, many enterprises were run like 'company towns', with a strong flavour of the paternalistic oppression imposed to curb collective action on the part of the workers.

Despite these and other formidable obstacles, sugar production is universally distinguished in this period by the increasing militancy of its workers. Expressions of worker resistance varied according to local circumstances from the age old strategies of burning cane fields, breaking tools, insubordination, absconding, to the full gamut of industrial tactics, including major strikes. Resistance arose in response to, among other things, poor conditions, the threat of disease, low wages and forms of payment (by truck or by ticket), and the refusal of most sugar producers to allow collective bargaining. In Australia and Cuba, sugar workers also struck to protect their jobs by forcing immigrant workers from the industry. But perhaps the most significant form of worker resistance found expression in the development of sugar workers' trade unions. Like so many of the changes in the industry throughout this period, the success with which trade unions developed and represented their members appears to have been highly differential. Where there was a long history of trade unionism and a broadly based and well developed economy, sugar workers became highly unionized. Indeed, in some cases, such as the Australian Workers Union, sugar workers' unions became extremely powerful and were able to exercise an influence not only over the industry as such, but within the trade union movement at large.

But, the experience of the Australian workers was not shared by their brother workers in most other sugar growing areas. This was apparent not only in such places as the Philippines or Peru, but also in the United States, where agricultural unionism in general, and attempts to organize sugar workers in particular were strongly and effectively resisted. But despite setbacks and defeats, the interwar period marked an important turning point for sugar industry workers. In many regions they had become a distinct force to reckon with. Much of this was due to broader political and social changes, but it was also closely related to changes in the labour process introduced by sugar producers to cope with the crisis. These changes displaced some workers but those who remained tended to have higher levels of skill and consequently greater potential bargaining power. In order to survive sugar producers had, therefore, sown the seeds which were gradually to undermine their absolute power and authority.

The role of the state

Not surprisingly the onset of WWI saw the role of the state in economic affairs increase immensely, and although there were attempts to return to laissez faire in the 1920s, the crisis of the 1930s saw the state having to take a more active role. A major factor conditioning this greater level of government intervention was the general economic and socio–political conditions of the period. The decades after 1914 can be characterized as a time of endemic crisis, dominated for much of the world by war and/or economic chaos. In some cases, most notably Nazi Germany, Salazar's Portugal and Primo de Rivera's Spain, these circumstances conditioned a move towards autarchy, with all that that implied for direct and intensive state intervention in economic and social life. More generally, however, states were quite justifiably concerned about the threat of war and moved to protect and encourage so called strategic commodities, which along with many other commodities such as oil and bauxite included sugar as an essential foodstuff. The chaos in the international payments system meant that states were also very sensitive to the foreign exchange issue. Import substitution was adopted almost universally, as was the imposition of tariffs and other protective measures to save foreign exchange. This also explains why commodities in surplus, including sugar, were sometimes dumped on the open free market at below the marginal cost, since although this may have represented a fiscal loss, the foreign exchange so earned was of considerable value.

The trend towards greater protection of the national industries, the negotiations over trade treaties and endless diplomatic efforts to regulate the world market meant that many governments were continually involved in the affairs of their sugar industries. Besides protecting them or arguing their case with other nations, the state sometimes took elaborate steps to encourage local sugar industries, either supporting them with subsidies and/or actually encouraging the formation of essentially new industries, as in Britain and India. There were many other forms of government help offered, including financial support, backing for research, the building of transport infrastructure, and perhaps most importantly, the exercise of police and military control over recalcitrant workforces. This last mentioned form of government support became particularly important for sugar producers when they they met resistance to their attempts to offset the crisis of the 1930s by cutting wages.

State protection and encouragement were not without strings, however, and tended to go hand in hand with much tighter state regulation of national sugar industries. Laws were enacted governing such matters as the disposition and use of cane and beet lands, relations between growers and millers and between millers and refiners. New authorities were established to purchase and distribute raw and refined sugars, to set the price of the raw material, and the wholesale, retail and export price of sugars. At the same time, land reform and the encouragement of peasant production was pursued in some countries and the enactment of appropriate labour

21

legislation including the establishment of conciliation and arbitration machinery occurred in others.

The question then arises as to whether the state was acting in the interests of particular groups in the producer nations. Fundamental structural changes in sugar production meant that generally the traditional sugar barons were a weakening force, especially in the larger more diverse economies where they constituted a small section of the society in any case. This trend was further complicated by the increasing mobility of capital in the period and the intervention in conflicts between local and foreign capital which strengthened the hand of the state. For all these reasons, the period saw an increase in the relative autonomy of the state vis-à-vis the old guard in sugar production, since the demands of this group were less homogeneous and influential than in the past, and legislatures were in a stronger position to dictate to the industry.

As the power of the sugar bourgeoisie declined governments increasingly favoured the interests of small producers and the working class. While there was no state which did not attempt to subdue the demands of workers and farmers, there was clearly far more ambivalence about wholesale repression. In the more advanced economies especially, both the working class and the petit bourgeoisie became better organized and more militant than hitherto, and the state demonstrated a willingness to accede to some of their demands. This was probably most obvious in Australia, where the state intervened between labour and capital across a very broad spectrum. States were increasingly prepared to accommodate workers' demands as a means to undermine the more general threat of working class insurrection, a spectre which haunted all ruling classes after 1917. Fear of widespread unrest also influenced attitudes towards the peasantry in that land reform, control over cane prices, the introduction of production quotas, and the development of peasant cooperatives all owed something to the state's perceived need to consolidate and win over the peasantry as a bulwark against Bolshevik revolution. This objective was complicated and overlaid in colonial territories by the fear of anti-colonialist revolution. While at a limited level domestic policies sought to accommodate that fear in the colonies, it also encouraged colonialist states to extend concessions and preference to the colonies to assure the producers of raw materials of the manifest advantages of remaining within the imperial fold.

How effective then, was the intervention of the state? Certainly, where it was strongest, sugar industries survived the traumas of the interwar period most successfully, although, as argued above, in many ways these traumas, particularly the apparent international problem of over-production, were due in large part to state intervention. However, given the economic and political conditions of the time it would seem that an active role for the state was unavoidable. For sugar this was nothing new, for as shown above, during the supposed 'Free Trade Era' much of the trade in sugar had been anything but 'Free'. But in the 1920s and 30s the breakdown of the international economy and increasing domestic social–economic and political tensions forced many states to extend their role.

Conclusion

As we have argued, the crisis of the interwar years wrought profound and irreversible changes in the international sugar economy. The industry's response was complicated both by the international framework within which it operated and by numerous factors which were local in origin. Beyond the mechanics and ramifications of the crisis of oversupply itself, the industry was deeply affected by the extraordinary turbulence of an era which included the trauma of war, revolutionary social and political upheaval, rapidly changing networks of international trade and, not least, the chaos in the international payments system. Sugar had always been a commodity characterized by two seemingly contradictory factors, in that it was one of the most international of goods while at the same time being the widespread object of exclusivist national economic policy. This latter tendency was greatly accentuated after WWI, as the world capitalist system deprived of the economic balance provided by Britain in the nineteenth century floundered under the weight of growing economic nationalism. All primary commodities and those who depended on them for their livelihoods suffered because of this. At the same time, as the essays in this book demonstrate, a broad range of overriding national and regional factors determined the specific character and form of the sugar producing countries' response to the crisis and the degree of success with which they emerged from it.

References

1. W. Arthur Lewis, ed., *Tropical Development 1880–1914. Studies in Economic Progress*, London, 1970, 19.

2. Figures from Noel Deerr, *The History of Sugar*, II, London, 1950, 490.

3. Manuel Moreno Fraginals, *The Sugarmill. The Socioeconomic Complex of Sugar in Cuba*, New York, 1976, 86.

4. Bill Albert, 'Causes del cambio technológico en la industria azucarera peruana, 1860–1940', in *Oro blanco y capitalismo*, ed., Horacio Crespo, Mexico, (forthcoming).

5. Ph.G. Chalmin, 'The Important Trends in Sugar Diplomacy before 1914', in Bill Albert and Adrian Graves, eds., *Crisis and Change in the International Sugar Economy 1860–1914*, Norwich and Edinburgh, 1984, 21–30.

6. Figures from, U.N., FAO, *The World Sugar Economy in Figures, 1880–1959*, Rome, 1960.

7. Bill Albert and Adrian Graves, 'Introduction', *Crisis and Change*, 1–2.

8. Manuel Moreno Fraginals, 'Plantation Economies and Societies in the Spanish Caribbean, 1860–1930', in Leslie Bethell, ed., *The Cambridge History of Latin America*, IV, c.1870–1930, Cambridge, 1985, 197–99.

9. Details on consumption taken from Sidney W. Mintz, *Sweetness and Power. The Place of Sugar in Modern History*, New York, 1985, chapters 3 and 4.

10. All figures on consumption from, FAO, *The World Sugar Economy*, 111–13.

11. Mintz, *Sweetness*, 149.

12. *Ibid.*, 186.

13. William H. Beveridge, *British Food Control*, London, 1928, 6.

14. Howard J. Gray, *War Time Control of Industry. The Experience of England*, New York, 1918, 169–70.

15. Beveridge, *Food*, 123.

16. Joshua Bernhardt, *Government Control of the Sugar Industry in the United States*, New York, 1920, 15. Also see Frank F. Anderson, *Prices of Sugar and Related Products*, War Industries Board, Washington D.C., 1919.

17. UN, ECLA, *Economic Survey of Latin America 1949*, New York, 1951, 91, 211, 271.

18. *The Economist*, Feb. 19, 1921, 359.

19. Prices from *Ibid.*

20. Alejandro García Alvarez, 'La Danza de los millones y sus consequences', paper presented at conference 'Crisis and Change in the International Sugar Economy 1914–1940 and the 1980s', Norwich, 1986.

21. League of Nations, *Sugar. Memoranda prepared for the Economic Committee*, Geneva, 1929, 32, 48–50.

22. FAO, *The World Sugar Economy*, 114. Stockpiles increased from about 5,000,000 metric tons in 1925 to 8,559,000 in 1930. The latter figure was equal to over a third of total production in that year.

23. Charles P. Kindleberger, *The World in Depression 1929–1939*, London, 1973, chapter 4.

24. V.P. Timoshenko, *World Agriculture and the Depression*, Ann Arbor, 1953, 122–3. Cited in , Kindleberger, *World in Depression*, 86.

25. International Sugar Council, *The World Sugar Economy. Structure and Policies*, II, London, 1963, 134.

26. B.C. Swerling, *International Control of Sugar, 1918–1941*, Stanford, 1949, 19.

27. A.G. Kenwood and A.L. Lougheed, *The Growth of the International Economy 1820–1960*, London, 1971, 175.

28. B.C. Swerling and V.I. Timoshenko, *The World's Sugar. Progress and Policy*, Stanford, 1957, 200–01.

29. *Ibid.*, 202–06. Also see Richard Lobdell, 'British Sugar Policy, 1919–1939', paper presented at conference 'Crisis and Change in the International Sugar Economy 1914–1940 and the 1980s', Norwich, 1986.

30. Swerling and Timoshenko, *World's Sugar*, 204.

31. International Sugar Council, *World Sugar*, 197.

32. U.S. Beet Sugar Association, *Concerning Sugar*, E–7–B & C.; International Sugar Council, *World Sugar*, 166–67; Swerling and Timoshenko, *World's Sugar*, 157–59.

33. International Sugar Council, *World Sugar*, 167.

34. Swerling and Timoshenko, *World's Sugar*, 159. For details on the impact of US policy on Cuba, see below chapter 8.

35. Murry R. Benedict and Oscar C. Stine, *The Agricultural Commodity Programs. Two Decades of Experience*, New York, 1956, 285.

36. *Ibid.*, 292–303. The Act was declared unconstitutional because of the use of a tax on processors to control production, but while some alterations were made the principles of the 1934 legislation were maintained in the Sugar Act of 1937.

37. Swerling, *International Control*, 40–2.

38. Unless otherwise noted, details on the Chadbourne Agreement and the 1937 International Sugar Agreement from *Ibid.*, 40–60.

39. This section is based primarily upon the following chapters and transcripts of the conference meetings.

40. Swerling and Timoshenko, *World's Sugar*, 149.

2

The German Beet–Sugar Industry and the Nazi Machtergreifung of 1933

John Perkins

Introduction

The longstanding debate on the links between German capitalism and National Socialism has been recently revised with a vengeance, following the publication in 1985 of Henry Ashby Turner Jr.'s *German Big Business and the Rise of Hitler*. This work of conservative scholarship seeks to exonerate the leadership of large–scale industry in general of having played a role in the Nazis coming to power in 1933. Those individual businessmen who became closely involved with National Socialism are considered by Turner to have acted out of personal eccentricity or because of a desire for self–advancement. The financial contributions to the Party's coffers by big business as a whole are viewed by him as a form of 'insurance', made in case of a Nazi victory at the polls and in order to support the more 'moderate' (i.e. pro–capitalist) elements in the Party leadership. Those scholars who have advocated the thesis that German business played a key role in the genesis of the Nazi *Machtergreifung* of 30th January 1933 are deemed by Turner to have been trying, for ideological reasons, to tarnish the public image of capitalism by linking it closely with the abominations committed by the Nazi regime.

Although some attention has been devoted by historians to the relationship between the leaders of heavy industry and the Nazi Party, there are few systematic studies of the interaction between individual industries and the political process during the Weimar era. Within the limited space allowed, this paper looks at one such important case study, the relationship between the Party and the beet–sugar industry. It cannot be said that this industry was typical of German industry as a whole, for there were many unique features, not the least of which was its agro–industrial character. Nonetheless, it was an industry of considerable importance, and one that, as a result of the formation of industrial combines in the early 1920s, certainly deserves to be classified as part of German big business. It is only on the basis of such studies that a clearer answer can be given to the points raised by Turner about this key issue in modern German historiography.

The sugar industry and Nazi political economy

In many respects a natural affinity might be expected to have existed between the National Socialist German Workers' Party (*NSDAP*) and the German beet–sugar industry (with the exception of the workers in the fields and the mills). First emerging during the French Revolutionary and Napoleonic wars, the beet–sugar industry was aptly described by one of its leading figures of the early 20th century as 'a genuine child of war', somewhat akin to the *NSDAP* that was spawned as a movement of *Frontkämpfer* returned from the Great War.[1] The leadership of both evinced a pronounced predilection for describing their everyday existence as a *Kampf* or struggle against their seemingly omnipotent but ultimately conquerable myriad of adversaries.[2] Moreover, the 'struggle' of both was essentially against the same forces.

From its early 19th century inception the beet–sugar industry, as an activity that was heavily dependent upon tariff protection and other means of state support against its cane rival, had been forcefully opposed by advocates of free trade: by the proponents of *Manchestertum*, which for the Nazis took the political form of the detested ideology of liberalism. For both 'Marxism' was a particular enemy. During the Weimar era this term was employed by Nazis and sugar–industry leaders alike to identify not only the ideology and protagonists of the German Communist Party (*KPD*) but also those of the German Socialist Party (the *SPD*).

The *KPD* was a particular target of attack for its anti–capitalist and anti–nationalist ideology. The *SPD* also represented a 'Fulfilment' policy in respect of the Treaty of Versailles, the *Versailles Diktat*, to which the beet–sugar industry was inclined to attribute the severe difficulties it experienced during the Weimar era, just as the Nazis were able to ascribe those of the entire nation. Additionally, the *SPD* was particularly opposed by the industry, and by the Nazis in their striving for the vote of peasant beet–growers, for its support of the 'consumer interest', in pressing for a domestic price of sugar that approximated more closely to the low and falling price on the world market.

In supporting a call for a 'national' sugar policy by the *Reichlandbund*, the main farmers' organization, the editor of a leading sugar–industry journal stated in April 1933 that:

As with the entire business community, so must the transformation of our domestic political situation, the 'National Revolution', be particularly welcomed by circles directly connected with the sugar industry. The Marxists have always seen the truly national German sugar industry as an enemy and, since the overthrow of the constitution [i.e. the adoption of the Weimar constitution in 1919] have left no opportunity untouched in making their powerful and evil policy felt by the industry. It is undeniable that beet–growers and the sugar industry would have come through the devastating effects of the world sugar crisis with much less damage and trouble if the Marxists, as the representatives of a purely consumer–oriented policy, had not seen to it that the supporters of the sugar industry were held back as much as possible.[3]

However, 'in future a fundamental change will occur, now that the sugar economy is liberated from the Marxist yoke'.[4]

Both the Nazis and the sugar industry emphasized their links with agrarian Germany. The former espoused a 'blood and soil' (*Blut und Boden*) ideology of the peasant embodying a Germanic racial and occupational ideal. The beet–sugar industry, in contrast to most others processing agricultural commodities, remained closely integrated within the agricultural sector. A large proportion of the beet continued to be drawn from growers who were also shareholders in the mills they supplied.[5] The cultivation of beet was labour–intensive and the sugar mills were necessarily punctiformly located in the countryside, on account of the high bulk to value ratio of the beet. The activity, therefore, was a counter–force to the rural depopulation caused by the rest of modern industrialization, which conservatives saw as sapping the 'lifeblood' of the nation and, in particular, diminishing the availability of healthy conscripts for the armed forces.

By the 1920s the industry produced an article that had become an important element in the diet of the masses, with the industry itself claiming that it had transformed sugar from a luxury to a necessity. A commodity that had once been entirely imported was now completely derived from domestic sources. Consequently, the beet–sugar industry was a classic example of national self–sufficiency, of the idea of autarchy that was so closely associated with the Nazis. That the industry had succeeded in its 'struggle' against cane–sugar, as a commodity produced by what the Nazis conceived to be inferior races in benign climatic conditions, was an added bonus in its favour. If a replacement for such an important 'colonial' product as cane–sugar could be developed at home, then the same could arguably apply to others; and was being applied in respect of synthetic rubber, fuel and textile fibres.

In arguing that the beet–sugar industry had a natural affinity for the Nazis, it is not intended to convey the impression that its leading figures were disproportionately represented in the membership of the *NSDAP*. In fact the political leanings of such figures was more towards the German National People's Party (*DNVP*) who were partners in Hitler's coalition formed in January 1933. With many originating from the *Junker* agrarian nobility, they tended to be monarchists, nationalists and conservatives, if not reactionaries. Nevertheless, they were emphatic about the rights of property and were supporters of free enterprise (within the bounds of a highly–protective tariff and other means of state support).

Typical of the political views of industry representatives were those espoused by Hans von Schieben, the son of a Silesian *Junker* and chairman of the Association of the Sugar Industry from 1930 until his death in 1932. In his report to the Association's annual general meeting of 1932, without contradiction from any of the assembled directors of sugar mills, he emphasised the difficulties for the industry stemming from 'our German state('s) ...fully unsuited parliamentary form of government'.[6] A year later August Winnig, a former head of the administration of East Prussia, found considerable support for his anti–democratic views amongst the audience of members of the Association. For Winnig, as for most of those present, the freedoms that had been fought for at the barricades of the 1848 Revolution – freedom of assembly, of the press, of trade and of movement – were no longer relevant. In particular, a democratically–elected parliament to which the government was responsible had been tried with the Weimar constitution and found wanting. As he put it: 'We have experienced the freedom of the people as the hegemony of parliament and we are fed up to the teeth with it'. In Winnig's view the word 'freedom' had acquired a new meaning with the 'National Revolution' of 1933. Instead of being applied to the individual, it now embraced the people (*Volk*) and its state as historical entities. Therefore, the task at hand was to free the people and the state from the 'shackles' imposed at Versailles: to 'overcome the situation of powerlessness' of the German state.[7]

Before and after the *Machtergreifung* of 1933, leading figures in the industry were apprehensive about the 'Socialistic' elements in the Nazi party and programme and, in particular, the ideas emanating from such Nazi leaders as Gottfried Feder (the Party's 'economic expert') the Strasser brothers and even Goebbels. These presented the prospect of enhanced state intervention in the operation of the capitalist economy in a form that was anathema to sugar millers. At the extraordinary general meeting of the Economic Association of the Sugar Industry on 21 March 1933, the chairman 'emphasised that the German sugar industry had complete trust in the new national government and expressed the wish that the government may remain strong and not get stuck halfway with its measures'. On the other hand, he was also emphatic that 'the freedom of the economy should not be interfered with'.[8] Here industry representatives were relieved by the appointment of *DNVP* representatives to the major portfolios concerned with the economy in Hitler's Cabinet and by that of the conservative Schacht, the rescuer of the currency in 1923, to the presidency of the central bank.[9]

While the sugar millers had reservations about aspects of Nazi ideology, the party had a strong appeal for two other and contrasting groups associated with the industry. On the one hand, support for the Nazis was pronounced among the chemists and technologists employed in the industry. On the other, National Socialism appealed to beet–growers and, especially, to peasants selling beet to mills of which they were not shareholders.

The 'modernist' aspect of Nazism attracted technologists, not least because it accorded them a high status as producers and as inventors of more efficient methods of production. Specifically appealing was the Nazis condemnation of speculative gains, of the activities of finance capital; whereby, in the words of one sugar–chemist, 'profit was made in the office rather than in the factory'.[10] In their capacity as the supervisors of processes of production, Nazism offered sugar technologists the prospect of an end to the conflict between labour and capital. Hitherto, apparently, the 'wage struggle' had predominated in the industry, had often made the technologist's supervisory function far from pleasant and had, in many instances, been the determining factor in the nature of the processes of production that were adopted.

The Nazis were particularly vigorous in their defence of peasant beet–growers, of the so–called *Kaufrübenbauten* ('sale–beet growers') who were not shareholders of the mills to which they sold their crops. This formed part of an overall concerted effort to attract the peasant vote, which followed the Party's electoral success in the countryside from 1930.[11] On the one hand, during the late Weimar period they opposed efforts by the Left parties to reduce the legal maximum price of sugar, on which that paid for beet depended. On the other hand, the Nazis supported the peasant growers in the conflict of interest that began to emerge within the industry as losses on exports mounted, which motivated the sugar mills to press for lower beet prices and restrictions on production.

Overall, sugar industrialists, technologists and beet growers were virtually unanimous in their acceptance, and mostly in their enthusiasm for the *Machtergreifung* of January 1933. This was reflected in the fact that the so–called 'co–ordination' (*Gleichsschaltung*) imposed by the Nazis left the leadership of the industry's organizations unchanged. For all concerned with the industry, other than the workers, the 'National Revolution' appeared to offer the prospect of resolving the crisis of the industry that had its origins in WWI and, not least, of eliminating the capital–labour conflict and the friction between beet–growers and sugar millers that threatened to destroy the harmony of interest upon which the industry had long prided itself.

The tariff and the sugar industry

The historical development of the beet–sugar industry in the 19th century may have furnished the Nazis with a classic example of the potential for autarchy. However, it was only during the early 1930s that the industry itself reluctantly came to accept the necessity to orientate itself exclusively towards the home market, just as it came to accept the need for some measure of state regulation. From the mid–1870s onwards Germany had emerged as a major exporter of sugar and by the eve of WWI only Cuba and Java supplied more sugar to the world market.[12]

After the war, and especially from the mid–1920s with the removal of state controls on the industry, the primary orientation was towards the recovery of the former position on the world sugar market. From 1925 Germany began once again to export sugar and by 1929–30 exports accounted for 15 per cent of production (as compared with two–fifths in 1913–14). However, Germany's reappearance in the international sugar market was associated with mounting losses. (Table 2.1)

Net exports were considerably less than the volumes shown in Table 2.1. Germany absorbed up to 100,000 tonnes of sugar a year from the Free State of Danzig, on payment of a nominal duty.[13] In addition foreign sugar was dumped on the German market with the downward slide of prices on the world market from late 1928, in consequence of the delayed response to the industry's demand for a drastic increase in the German tariff.[14]

Apart from the desire for protection against dumping, the industry argued for tariff increases on sugar to facilitate an increase of the domestic price, through co–operative marketing, in order to meet the losses on exports and to compensate for the loss of earnings during the 'Sugar Famine' of the early postwar years.[15] At that time, as a result of the drastic wartime reduction of productivity capacity and government control of the price of sugar on the domestic market, German producers had been unable to take advantage of the high prices ruling on the world market.

In July 1927 a favourable majority in the *Reichstag* enabled the duty on imported sugar to be increased by 50 per cent, to *RM* 150 per tonne on the refined product. In December 1928 the tariff was further increased, to *RM* 250 on refined sugar: or 25 per cent above the high level prevailing before the Brussels Convention of 1902. Finally, in March 1930, the duty was raised to the absolutely prohibitive level of *RM* 320 per tonne on refined sugar, at a time when it was selling on the world market for about *RM* 100 per tonne.[16]

Table 2.1. *German Sugar Exports 1925/26 to 1929/30.*

Season	Exports (tonnes raw value)	% of production	Losses (millions of marks)	On Exports (marks per ton of beet)
1925–26	135,000	8	9.4	1.00
1926–27	195,000	12	16.7	1.60
1927–28	130,000	8	14.6	1.40
1928–29	235,000	13	40.1	3.50
1929–30	290,000	15	62.3	5.20

Sources: K. Sewering, *Zuckerindustrie und Zuckerhandel in Deutschland*, Stuttgart, 1935, 18; *Centralblatt der Zuckerindustrie*, 22 Feb. 1930.

In order to limit the discordance between free–market and domestic sugar prices resulting from tariff increases from December 1928, and as a concession to the 'consumer interest' represented by the Left in the *Reichstag*, provision was made for the duty to be reduced by *RM* 100 per tonne when the monthly average price on the Magdeburg Exchange exceeded *RM* 420 a tonne. Subsequently, from June 1929, this arrangement was modified to ensure an even flow of sugar onto the market by compensating producers for storage costs, by means of raising the maximum allowable price on the Madgeburg Exchange by three marks a tonne per month from January to September (to a maximum of *RM* 470 in the latter month).[17]

German participation in the international sugar economy from the mid–1920s was only possible through the formation of a central organization to apportion the mounting losses amongst individual producers. Subsequently, with the tariff revision of December 1928 this organization, initially the Export Association of German Sugar mills and later the Association for Consumption–Sugar Distribution, took over the task of determining the volumes and timing of the mills' releases of sugar onto the market. By such means seasonal price fluctuations were minimized and the price on the Magdeburg Exchange was maintained as close as possible to the permissible maximum before a reduction of the duty.[18]

Exports and the home market

From 1931 to 1933 the industry's difficulties with exports were compounded by ones emerging on the home market. For the 1930–31 season domestic sugar sales increased by a mere 0.9 per cent, as compared with an historical average of three per cent per annum. This reflected the rapid rise in unemployment and suggests that the income–elasticity of demand for sugar remained quite high. The doubling of the excise duty on sugar from 5 June 1931, as part of the parcel of measures adopted by the Brüning government in response to the banking and foreign exchange crisis, contributed significantly to a 12.5 per cent fall in sales for 1931–32. Declining domestic consumption of sugar coincided with an exceptionally high level of output in the 1930–31 season, showing a 29 per cent increase in production over 1929–30.[19]

In the face of a worsening crisis, the 'free rider' problem took on unbearable proportions. More and more individual firms became unwilling to share the burden of losses involved in exporting sugar and in delaying deliveries to the home market in order to even out seasonal price fluctuations. The growers were in revolt at increasing pressures from the mills to restrict production and reduce beet prices. As a result the wheel of opinion in the industry turned full circle. Outright opponents of the 'regulated economy' (*Zwangswirtschaft*) of the war and early postwar years had now become convinced of the need for government intervention.

The Ministry of Agriculture responded with a decree of 27 March 1931, by which mills and refineries were compelled to join an Economic Association of the German Sugar Industry, the forerunner of the Economic Group Sugar Industry of the Nazi era. The Association was charged with the task of balancing supply and demand on the home market, promoting increased domestic consumption and allocating export quotas. Such a legally constituted body was also necessary to enable the German industry to enter into international agreements to combat the sugar crisis, such as the Chadbourne Agreement of 9 May 1931.[20]

Under the terms of the Chadbourne Agreement, Germany was allocated an export quota of 500,000 tonnes for 1930–31, declining thereafter to 350,000 tonnes for 1931–32 and to 300,000 tonnes per annum for the remaining three years of the agreement. In practice, after reaching 83 per cent of the quota for 1930–31, only 99,000 tonnes were exported during 1931–32. Losses on exports were such that the German industry requested a reduction of its quota for 1932–33 to 200,000 tonnes; a figure that was not met. For 1933–34 the industry declined to fulfil its quota of 75,000 tonnes, and Germany finally ceased to participate in the international sugar economy.[21]

The failure of the industry to meet its export quotas combined with the abnormally high output of the 1930–31 season and declining domestic consumption, necessitated a drastic reduction of production. This was made possible by the legal powers granted to the Economic Association. A basic output figure, the *Grundkontingent*, was fixed at 2.05 million tonnes, or about 25 per cent below actual production in 1930–31. A formula was devised to apportion the *Grundkontingent* between the various regions, with individual mills being allocated quotas that were transferable where the recipient agreed to accept beet from growers attached to the transferer.[22]

On account of wide annual yield variations, beet delivery quotas were fixed by the mills on the basis of the average of the three seasons prior to 1930–31. In effect this arrangement gave the 'Sale–Beet Growers' (*Kaufrübenbauern*) a right to deliver beet to a mill. At the same time, such growers lost their former right to sell to the mill offering the best terms and, where mills experienced particular difficulties, they could reduce deliveries from such growers by up to 10 per cent.[23] These developments, and the continuing low price paid for beet, were sources of considerable dissatisfaction amongst the 'Sale–Beet Growers'.

Actual production quotas were supposed to be fixed each year by the Economic Association, as a proportion of the *Grundkontingent*, in time for the mills to allocate beet quotas to growers. The initial output figure set for 1931–32, at 80 per cent of the *Grundkontingent* was revised to 65 per cent (or about 53 per cent of output in 1930–31) on 27 August 1931, or shortly before the beet was due to be harvested. For the 1932–33 season it was initially fixed at 63 per cent and finally, on 10 August, raised to 67 per cent. In addition to uncertainty as to the final quotas, variations in the sugar–yield of beet occasioned mills to allocate beet quotas in excess of those required with a normal harvest. Growers in turn tended to grow even more beet for fear of not being able to meet their quotas due to a bad harvest.[24]

Conclusion

The catastrophic experience of the German beet–sugar industry during the Great Depression of 1929–1933 followed a decade and a half of crisis, which commenced with the outbreak of WWI. The recovery of production during the years 1925 to 1928 was not accompanied by a significant improvement in profitability, in consequence of rising losses on exports and rising costs of production. The ensuing drastic decline of prices on the free market inevitably enforced an orientation towards complete dependence upon the home market. Because of this, however, thanks in part to Chancellor Brüning's strict deflationary policy–response to the Great Depression, and the confusion in the industry wrought by the quota system for producers, the domestic market also began to approach the point of collapse. All in all, therefore, it is far from surprising that industry leaders enthusiastically welcomed the 'National Revolution' of January 1933. The stability and order it seemed to offer for the political sphere promised to be reflected in the state of the economy.

References

1. B. Brukner, *Zucker und Zuckerrübe im Weltkrieg* Berlin, 1916, 5.

2. See, for example, *Ibid.*, 80; G. Tannenberg, *Der Kampf um den Zucker*, Leipzig, 1944, *passim.*

3. *Centralblatt für die Zuckerindustrie (CBZI)*, 1 April 1933.

4. *Ibid.*

5. See J. Perkins, 'The Political Economy of Sugar Beet in Imperial Germany', in B. Albert & A. Graves eds., *Crisis and Change in the International Sugar Economy, 1860–1914*, Norwich & Edinburgh, 1984, 39.

6. *Zeitschrift des Vereins der Deutschen Zuckerindustrie (ZVDZI)*, LXXXII, 1932, *Allgemeine Teil*, 113–6

7. *ZVDZI*, LXXXIII, 1933, *Allgemeine Teil*, 256–7.

8. *CBZI*, 25 March 1933.

9. *ZVDZI*, LXXXIII, 1933, *Allgemeine Teil*, 144: Report of Association Directors for 1932.

10. *Ibid.*

11. See I. Farr, '"Tradition" and the Peasantry; on the Modern Historiography of Rural Germany', in R.J. Evans & W.R. Lee eds., *The German Peasantry*, London, 1986, 7.

12. See Perkins, 'Political Economy of Sugar Beet', 31–46; 'The Agricultural Revolution in Germany, 1850–1914', *Journal of European Economic History*, X, 1, 1981, esp. 81–4.

13. It was a source of complaint by the industry that a 'national obligation' as regards Danzig had to be effectively borne by sugar producers.

14. See K. Sewering, *Zuckerindustrie und Zuckerhandel in Deutschland*, Stuttgart, 1933, 13; R. Knuth, *Die gegenwärtige Weltmarktslage für Zucker und der Gedanke einer neuen Zuckerkonvention*, Diss.: University of Rostock, 1928, 37.

15. *ZVDZI*, LXXXVI, 1926, *Allgemeine Teil*, 156.

16. *Ibid.*

17. Sewering, *Zuckerindustrie*, 15, 17.

18. *Ibid.*, 166; H. Spohr, *Die Marktordnung der deutschen Zuckerwirtschaft unter Berücksichtigung der nationalsozialistischen Marktregelung vom 11. 11. 1934*, Diss.: University of Breslau, 1936, 3.

19. *ZVDZI*, LXXXIII, 1933, *Allgemeine Teil*, 87.

20. *Ibid.*

21. Spohr, *Die Marktregelung*, 33–6.

22. *Ibid.*, 56–7; *ZVDZI*, LXXXII, 1932, *Allgemeine Teil*, 161.

23. Spohr, *Die Marktregelung*, 60.

24. *Ibid.*, 61; *CBZI*, 24 June 1933.

3

The Crisis of the Beet Sugar Industry in Czechoslovakia[1]

Frantisek Dudek

Problems and structure 1920–1940

In interwar Czechoslovakia food processing was the third most important industry after metal–working and textiles.[2] The former accounted for roughly 20 per cent of industrial output from the first stage of capitalist industrialization in the mid–nineteenth century until the economic crisis of the 1930s. The continued strong position of food processing, especially of the sugar industry, which accounted for about one half of the food industry's output,[3] was the result of a relatively delayed, slower, and uneven capitalist development in the Czech and Slovak territories (the historical Czech Lands, i.e. Bohemia and Moravia, and part of Silesia which formed the economically most advanced western part of the Czechoslovak Republic since its founding in 1918) incorporated in the fairly extensive, but economically weak, market of the multi–national Habsburg (Austro–Hungarian) monarchy.

During the first half of the 1920s the Czechoslovakian sugar industry benefited from the postwar price boom, and by the campaign of 1925/26 production and exports had surpassed prewar totals. But, the exceptional profits made were not utilized sufficiently to modernize production, especially in the Czech Lands where there were still many small plants and those with outdated equipment. The industry was not, therefore, in a good position to respond to the 1925 price collapse and subsequent crisis in the international market.

Until the mid–1920s Czechoslovakia accounted for 15–17 per cent of of world sugar beet production, two–thirds of which was exported. This was reason enough for the country's sugar producers to take part in the Cuban inspired Tarafa negotiations with their counterparts from Cuba, Germany and Poland in Paris at the end of 1927. The object was to cut production and exports so as to improve world prices. Czechoslovakia agreed to reduce exports by 6 per cent in the 1928/29 campaign, and despite the fact that the agreement was never concluded, the country did restrict its sugar production to some extent in any case.

Czechoslovakia took part in the next abortive international conference in Brussels in 1929, and the somewhat more successful Chadbourne Agreement on export restriction, concluded in 1931. When this latter accord also failed to curb the sugar crisis, yet another meeting was called in London in 1937. Under the International Sugar Agreement which came out of this conference Czechoslovakia remained the single largest beet sugar exporter, but by this time, because of development in the USA, USSR, and other European countries, its share of world exports had fallen from 17 per cent in 1925 to a mere 7 per cent, and its share of output had been halved, from 6 to 3 per cent.

From the 1932/33 campaign exports as a proportion of output fell from about two–thirds to one–third as a result of the restrictions imposed on production for foreign markets. Although the amount had risen to 50 per cent by 1936/37, it remained below its earlier levels by the outbreak of war. The crisis was also reflected in the substantive changes in the direction of Czechoslovak sugar exports. By the early 1930s sugar had been pushed out of its original markets in Central and Western Europe and to a large extent in south–eastern Europe. New outlets were sought, although with difficulty, in Scandinavia and the Baltic States of Finland, Latvia and Lithuania. But in all these markets there was strong competition from cheap cane sugar from British refineries, as well as Polish and Soviet sugar. In the 1930s, Czechoslovakian sugar was also being exported to the Near and Middle East, North Africa and parts of South America. In many of these markets it again had to compete with British as well as Javanese sugars.

Czechoslovakia also had other export problems. Although the the water route to Hamburg along the Elbe, which carried about half the country's sugar exports, was relatively inexpensive, being landlocked did impose extra transport costs. During the crisis years these costs rose to as much as a third of the steadily falling price of exported sugar. Things were made more difficult for Czechoslovak producers because of their reliance on foreign middle men. If all this was not enough, the only way the country could export its sugar was by dumping it below cost onto foreign markets. With unrefined sugar this practice began in the middle of the 1920s and by the 1934/35 campaign the export price covered only 28 per cent of production costs. Subsequently it increased, but by 1937/38 was still only 50 per cent. The price of refined sugar exports also fell below costs from the end of the 1920s, but not to the same extent as that for raws. Throughout the crisis period Czechoslovakian sugar exports were highly unprofitable, and this position was made more difficult by the fact that the country remained on the gold standard after 1929. It was only after the worst point of the depression had been passed that the government devalued the currency, in 1934 and again in 1936. The industry only avoided complete collapse through a combination of state subsidies and increased protection of the domestic market, the latter resulting in high sugar prices for Czech consumers.

The sugar crisis was quickly reflected in decline in the relative significance of the industry in the country's foreign trade. Up until 1926 it had made an important contribution to Czechoslovakia's favourable trade balance, but from this point its role declined, from 12 to 14 per cent of export value to only 1.7 per cent in 1935/36.

The crisis described above had a powerful impact on the structure of the Czechoslovakian sugar industry. In the mid–1920s there were 144 sugar plants in the Czech Lands producing an average of 8,655 metric tons of sugar per plant and 11 such enterprises in Slovakia, with an average yearly output of 23,533 tons. Because of the remunerative prices up to this time little progress had been made either in modernization or consolidation in the sugar industry in the Czech Lands. However, in the more hostile climate between 1928 and 1935, 42 plants with less than 5,000 ton capacity disappeared. But for those left the radical cutbacks in output meant that capacity tended to be chronically under–utilized and this raised fixed costs. In the end, although there was some change, the rationalization of production did not correspond to the scale of the crisis which the industry faced.

The sugar cartel

The reasons for this failure must be sought in the political economy of sugar which emerged in these years. One of the bases for this was the founding in 1926–27 of a sugar cartel. This in turn was the result of a struggle for control of sugar industry policy, for example, quotas for the domestic market or high monopoly profits, between the industrial and agrarian groups. The economically weaker agrarians could not compete effectively against the financial power of the large commercial banks when they took over sugar operations.[4] That is why the Czech agrarian bourgeoisie interested in sugar had to rely on their political muscle to try and influence the government. Under the auspices of the presidium of the Republican (Agrarian) Party, and with the tacit support of the agrarian members of the government, headed by the prime minister Svehla, support was given for the setting up of a cartel. This was done on condition that the large sugar–beet growers, who in the 1920s controlled 34 per cent of the joint–stock capital of the sugar plants, would be represented. The agreement forbade the sale of sugar over a fixed quota on the domestic market, as well as the building of new sugar plants. The cartel further controlled sugar supply to the domestic market, determined its grinding price and regulated the production and trade relations between the raw sugar plants and the refineries.

One of the cartel's most significant immediate actions was to increase the domestic price of sugar by 26 per cent by 1928. It also moved to lower the wages of sugar plant workers. After an unsuccessful attempt at the beginning of the 1926–27 campaign, it finally managed in 1931 to cut wages by abolishing exceptional bonuses and campaign payments. The cartel also worked against small and medium sized beet growers (up to 20 hectares). This began in 1926 with a 25 per cent drop in the buying price of sugar beet. The price had thus been reduced to the level of these farmers' costs (about 15 Czechoslovak crowns per metric hundredweight of beet).

As part of the 1927 international negotiations in Paris, the sugar cartel and the beet growers' organization formed a joint Commission to Protect Czechoslovakian Sugar Beet Growers and the Sugar Industry. This new group was to take care of international dealings with other countries' industries and lead efforts to win over the public, but more importantly the government. In the course of its widespread propaganda campaign the state was asked to decrease the relatively high excise tax on sugar so that its domestic price could be increased and, therefore, also the price of sugar beet. In this way the Commission wanted to underwrite export dumping and maintain the price of sugar beet above most growers' costs. After the Minister of Finance had resigned following the unpopular increase in the price of sugar at the end of 1928, the cartel forced the government to decrease the domestic sugar turnover tax, to pass the law on reimbursing commercial taxes and to promise to rebate the taxes paid by the cartel to the amount of 45 million Czechoslovak crowns over a period of three campaigns. In return for these concessions the cartel and the sugar beet growers drew up a joint economic and financial plan to restore the soundness of sugar–beet growing and the sugar industry over the campaign years 1928/29 to 1930/31. Moreover, the agricultural bourgeoisie managed to obtain a proportion of the state funds to maintain sugar beet prices and to aid the so–called agricultural sugar plants it controlled.[5] The remaining money was used by the cartel to subsidize refined sugar exports and restrict raw sugar exports. The agricultural sugar plants in fact received subsidies of 10 million crowns each year for over a ten year period. The total subsidy during the three campaigns mentioned amounted to 645 million crowns.

Until 1930, sugar beet prices were maintained at the cost of production level of small and medium growers, who supplied half of all beet. At the same time the fall in sugar exports continued to fall. This led to a decrease in sugar beet price despite its being subsidized. The result was a call from the sugar beet growers to restrict acreage as the only way of preventing further price collapse. The sugar plant owners opposed this move as they feared this would lead to a closer of plants and a forced concentration of production.

A change in approach came only after the plant owners and beet growers came together to sign the 1931 International Sugar Agreement. In that year the cartel and the sugar beet growers ordered a uniform restriction of beet and sugar output of about 20 per cent. Both the growers and the plants were assigned production quotas, the latter also having to observe export quotas. These plans were given full government backing. The state introduced export controls so as to convince all sugar plants to agree to the production restrictions. A law was passed forbidding the building of new sugar plants and another law prohibited the reopening of plants closed in the three years before 1932. By the Minister of Trade issuing export licenses, the authorities took a key part in the practical implementation of the International Sugar Agreement.

The results of all these measures was an almost 30 per cent drop in beet acreage in the campaign of 1931/32, together with a fall of 27 per cent in both sugar production and exports. But, even though world production and exports also fell at this time, prices continued to fall and this led Czechoslovakian sugar plant owners to ask for a further cut in their raw material prices. This was difficult, for even including the subsidy, beet prices were just above cost incurred by the large growers and below those of the small farmers. The response of the beet growers' association was to force yet another cut in acreage and sugar output for the 1932/33 campaign. Quotas were cut by a further 20 per cent, far beyond the obligations entered into by the country in the 1931 Agreement. Thus, over a two year period beet and sugar production had been slashed by more than 40 per cent, partly to conform with international obligations, but also at the same time to alter the unfavourable ratio of unprofitable sugar exports to profitable domestic sales. This was achieved. After the second reduction exports were down by 60 per cent, with the proportion of output exported falling from two–thirds to one–third. The impact on the country's industry was, however, uneven. On the whole the Czech agrarian bourgeoisie benefited, while those in Slovakia were badly affected by the reductions in acreage, production and exports. In this latter region beet prices were extraordinarily low and sugar exports had practically collapsed. Furthermore, although in Slovakia sugar beet growing was in the hands of large farmers they had less political influence than the growers in the Czech Lands. The latter provided important electoral support for the ruling agrarian Republican Party. Overall production was maintained at the new lower level until 1937, when a slight increase was allowed as a tactical measure to improve the country's bargaining position at the international conference held that year in London.

During the campaigns of 1931/32 to 1933/34, the sugar cartel introduced a new system for determining sugar beet prices. These were now to vary in line with changes in the world sugar price. As this continued to fall the consequence, in spite of radical restrictions and costly subsidies provided by both the high price of domestic sugar and state funds, was that the price of sugar beet was barely maintained. Throughout the 1930s it tended to be below small farmers costs and just above those of the larger growers. The reduced volume of beet production, therefore, increasingly passed into the hands of the agricultural bourgeoisie (i.e. large farmers with more than 20 hectares of land) who, together those controlling the sugar plants, profited at the expense of the smaller beet farmers.

By undermining the more marginal beet sugar producer, the sugar crisis and the way it was handled in Czechoslovakia reduced standards of living in the Czech Lands, where small growers were predominant. The smaller farms were forced to restrict plantings to a much greater degree than they were technically obliged to do. This was partly due to the fact that the principle of uniformly controlled restrictions, which were set regardless of acreage, and pushed through by the large sugar beet growers, had a more severe impact on the smaller farmers. An important consequence of this was that instead of sugar beet they turned instead to corn, potatoes, fodder beet, and other products.

The increasing production of grains had a knock–on affect with respect to the country's wider trade relations. It led, for example to the upsetting of trade with the neighbouring grain exporting Balkan states, the traditional market for Czechoslovakian sugar as well as industrial goods. This in turn had an important repercussions on the country's foreign policy, especially the Little Entente, designed in part to foster closer trading ties between Yugoslavia, Rumania, and Czechoslovakia. This is why the industrial fraction of the bourgeoisie as well as the government were concerned in maintaining the specialized nature of agricultural production, with a large share continuing to be taken by sugar beet.[6]

The sugar–plant owners too were opposed to excessive limitation of beet acreages because they wanted to be able to renew their exports quickly as soon as the world sugar market picked up again. However, the large beet growers were only willing to make concessions if sufficient profits were guaranteed. These issues became a major preoccupation of state economic policy in the 1930s, causing serious controversies in the government. They also led to disputes among the leaders of the influential Republican Party, in which a change of generations was taking place. While they were able to continue winning support for their sugar policies, the older generation of the agricultural bourgeoisie, associated with the interests of the sugar industry generally, was giving way under the increasing pressure of the crisis to a more conservative group which supported the interests of the corn and potato growers. These changes had a negative impact on political developments in Czechoslovakia during the second half of the

1930s, particularly with regard to increasingly close economic and political contacts with the Nazi government in Germany.

After the second round of production restrictions in 1932 sugar exports continued to slump, and, because of the impact of the general economic crisis, domestic consumption also fell. This led to a major new departure for the sugar industry, the production of denatured untaxed sugar as animal feed. It began to be produced, with the approval of the Ministry of Finance, after the first cutbacks in 1931, from sugar produced in excess of the stipulated quotas. But the demand for this new product did not increase substantially until the bad fodder crop harvest of 1934 and the lowering by the cartel of the price of denatured sugar to 65 crowns per metric hundredweight (the price of the equivalent amount of refined sugar was 555 crowns). In rural areas seriously affected by the economic crisis the denatured sugar, because of its cheapness, was often used as a foodstuff by the poor. Because of this its distribution was forbidden for a time.

Another method of using excess beet and raw sugar was provided for by a 1932 law which made it compulsory to mix spirit with petrol. This had been pushed through by the agricultural bourgeoisie, and it allowed the manufacture of spirit not only from molasses but also from beet and sugar. However, neither use as fodder or as a partial petrol substitute proved very important for the industry, only a maximum of 4 per cent of both sugar beet and raw sugar being taken for these two products. The attempts at alternative use came up against the more general economic crisis which also hit other agricultural commodities, with low prices being obtained for fodder, cereals and potatoes. There were other attempts at by-product manufacture, including the use of sugar in the making of rubber, explosives, edible fats, wood dye, and so on, but none of these made a significant contribution to reducing the crisis.

Until the destruction of the Czechoslovak Republic in 1938, the sugar cartel managed to keep the domestic price of sugar at the October 1928 level. With the support of the government it resisted the strong pressure of dissatisfied consumers, who called for a drop in prices after the unprofitable exports had been cut back. In the boisterous system of pluralistic bourgeois parliamentarianism, with Socialists participating in the government coalition and a large group of Communists in the parliament, the sensitive question of the domestic sugar price became important in appealing to voters. However, nothing but promises were made, as the entire affair became the subject of complex negotiations. But for the historian, at least, the parliamentary activity does reveal some interesting facts. A parliamentary subcommittee was established in 1936, to deal with the possibility of lowering sugar prices. The evidence it collected indicated that the costs of one kilogram of refined sugar, which sold at an average retail price of 6.30 crowns, was only 4 crowns including tax and normal profit. The cartel was, therefore, reaping a monopoly profit of about

900,000,000 crowns per year. In addition it was receiving roughly another 100,000,000 crowns in the form of tax rebates and other advantages, in order to underwrite exports and maintain sugar beet prices.

Since exports fell during the thirties, so did the estimated losses attributable to the subsidy. For example, in 1931 the total loss was about 600,000,000 crowns, but by 1935 this had fallen to only 137,000,000. The seemingly paradoxical tendency of the cartel's profits to increase during the crisis was also due to the decrease in costs at the sugar plants, achieved partially by the concentration and rationalization of production, but also because both wages and beet prices were reduced. As a result of these factors, together with production restrictions in 1931 and 1932, profits made by the sugar plant owners and the large sugar beet growers increased from 650,000,000 crowns in 1932 to 750,000,000 by 1935.

The well organized monopolization of the Czechoslovakian sugar industry, supported by state intervention, prevented the total collapse of unprofitable sugar exports and offered high profits to producers. A clear measure of this is given by the fact that before output was cut by one-third in the 1930s no outflow of capital took place in sugar as it did in most other light consumer goods industries. The average annual index of sugar enterprise stock was back at its 1929 level only two years after the 1932 panic on the Prague stock exchange. Quotations rose from this point reaching nearly double their original value by 1937. The increase in profits is also borne out by the published shareholder reports of the various joint-stock companies.[7]

To understand the importance of the sugar cartel within the Czechoslovak economy during the crisis years it is necessary to consider the political role and the strength of the agrarian interests. The cartel was a major tool in allowing the agrarian group of Czech capitalists to secure their position within the economy under the auspices of the most influential political party. The politically dominant Czech agricultural bourgeoisie continued to move into the sugar industry during the 1930s, not by directly controlling the majority of the sugar industry but by gaining the political support of the government. This became apparent especially during the negotiations for a new cartel agreement in 1937.

The preservation of the sugar cartel was an absolute necessity if high profits were to continue to be earned from a controlled domestic market. In the drawn out negotiations between the industrial and agricultural bourgeoisie, which took place well before the 10 year cartel agreement expired, the latter were able to obtain large domestic quotas for its plants. The renewal of the cartel was supported by the government, which, among other things, wanted it to become the executive arm of the International Sugar Agreement of 1937. Official backing came despite the fact that under

Table 3.1. The Czechoslovakian Sugar Industry 1925–1938.

Campaigns	Plants	Area in beet (000 ha)	Beet milled (000 t)	Sugar Output (000 t)	Sugar exports (000t)	Consumption total (000t)	Consumption per head (kg)	Surplus (000mt)
1925/26	166	312	8826	1510	1080	408	28.7	19
1926/27	163	258	6233	1042	708	370	25.5	-18*
1927/28	162	281	7476	1254	813	393	27.2	30
1928/29	160	250	5988	1057	662	407	27.8	18
1929/30	148	227	5553	1035	606	405	27.7	43
1930/31	146	237	6758	1143	570	401	27.2	215
1931/32	139	179	4429	815	505	396	26.4	129
1932/33	126	138	3579	634	280	399	24.8	84
1933/34	121	139	2811	517	166	401	24.7	34
1934/35	119	145	3777	638	222	409	24.2	41
1935/36	119	146	3353	571	168	410	25.4	35
1936/37	119	149	4066	728	315	424	25.6	24
1937/38	118	167	4666	755	340	-	-	-

Note: * negative reserve indicates overconsumption for next campaign .
Source: Kalendar cukrovarnicky, XVII, 1938, 576, 578-81.

the Agreement Czechoslovakia had to decrease its beet acreage below the already lowered levels achieved in 1932.

In the reformed cartel the sugar plants owned by the agricultural bourgeoisie achieved a distinct predominance as a result of the backing given by the leaders of the Republican Party and its cabinet ministers. Although this group held only one quarter of the total sugar plant's joint-stock capital in the mid–1930s, their share of the domestic raw sugar quota was increased from 33 per cent to 44 per cent and that for refined sugar from 15 per cent to 38 per cent.

The negative impact these changes had on the other groups in the industry was soon eliminated by the wartime sugar boom. As domestic demand increased the price of sugar beet rose above 20 crowns per hundredweight (the last time the price was at this level was during the campaign of 1924/25) and quotas were lifted.

Conclusion

The experience of the Czechoslovak sugar industry during the prolonged sugar crisis years is important for a number of reasons. The fact that the country was one of the largest producers and exporters of sugar in the world meant that her policies could not help but have a major impact on the world market. Although it is difficult to gage precisely the international effects of Czechoslovakian sugar policy, the reduction of both output and exports must have eased, at least to some extent, pressure on prices. However, at the same time the dumping in foreign markets of artificially low priced sugar and the protection of her own market worked against world recovery. Internal policies also had contradictory effects. By guaranteeing high profits for favoured producers in the country, the Czechoslovak government, under the strong influence of the Republican Party, maintained the sugar industry at the expense of the consumers, the industry's workers, the small beet growers, and also of the industry's own modernization, a modernization seriously hindered for decades by the monopoly power wielded by the agricultural bourgeoisie through the sugar cartel, which had existed off and on in various forms since 1890. Finally, events in Czechoslovakia illustrate two closely related issues which were of major significance for the international sugar economy in the interwar years. The first was the central importance for the survival of sugar production in most countries of the sugar interests being able to wield political power and with it to obtain state support. It was this support which in turn contributed to making these industries unresponsive to world market pressures and so helped to deepen and prolong the international crisis.

References

1. Unless otherwise stated, this study is based on the material in the author's, *Monopolizace cukrovarnictví v ceskych zemích do roku 1938,* (Monopolization of the sugar industry in the Czech Lands up to 1938), Prague, 1985. With respect to the causes, events, and consequences of the Great Depression on Czechoslovakia, the author has drawn on the most recent paper by Vlastislav Lacina, 'Velká hospodárská krize v Ceskoslovensku', ('The Great Depression in Czechoslovakia 1929–1934'), Prague, 1984.

2. V. Lacina, 'The Structure of Czechoslovak Industries and its Changes in the Thirties of the 20th Century', *Economic History,* (Prague), 2, 1978, 55–96.

3. F. Dudek, 'Utvárení základu zemedelskoprumyslového komplexu v procesu kapitalistické industrializace Ceskych zemí', (Forming the foundations of the agro–industrial complex in the process of capitalist industrialization in the Czech Lands), *Economic History,* (Prague), 9, 1982, 7–63.

4. These problems are described in detail in, F. Dudek, 'The sugar cartel and the penetration of agrarian capital into the sugar industry of Czechoslovakia 1918–1927', *Historica,* (Prague), XXI.

5. We denote the plants whose shares were entirely or in part dependent on the consignment of a determined amount of sugar beet and hence were in the hands of the beet growers by the term agricultural sugar plants. For details see, Ibid., 105–36.

6. In the fertile plains of the Czech Lands there were two regions where sugar beet accounted for between 20 per cent and 30 per cent of arable land. In Bohemia, from the 1870s and Moravia, from the 1880s, sugar beet covered more than 15 per cent of arable land in the regions where it was grown. No other European regions could claim the same high proportion. See B. Albert and A. Graves, eds., *Crisis and Change in the International Sugar Economy 1860–1914,* Norwich and Edinburgh, 1984, 8a; V. Lacina, *Krize ceskoslovenského zemedelství 1928–1934,* (The Crisis of Czechoslovak Agriculture 1928–1934), Prague, 1974, 45–70; F. Dudek, *Monopolizace,* cartogramm 2, 193.

7. *Annual Report of the Czechoslovak National Bank for 1937,* Prague, 1938, 76; *Reports of the State Statistical Bureau,* 15, 1934, 128–30, 964–65; 16, 1935, 152–54, 1138–43; 18, 1937, 99–102, 744–99; *Reports of the Czechoslovak National Bank,* 1937, no.131, 485–90, 1938, no.140, 324–28.

4

The Spanish Sugar Industry, 1914 – 1936

Manuel Martín Rodríguez

The immediate effects of WWI on the Spanish sugar industry

The high protectionist barriers erected against competition from abroad, together with other interventionist policies, had impeded the levelling out of Spanish and foreign prices by normal market forces. Rivalry between the 'Sociedad General Azucarera', the most important sugar company in the country set up in 1904 at a moment of grave crisis for the industry[1], and the independent factories had given rise to a fierce struggle to corner the sugar beet market, the final result of which was to pass on all the benefits gained from the tariff barriers to the landowners and leave the sugar manufacturers in dire straights. Furthermore, the high prices of beet occasioned by this war did nothing to encourage the landowners to introduce new techniques into their cultivation methods, which eventually resulted in further lack of efficiency and concomitant price rises.[2]

In spite of the structural problems, however, as with other food products, the Spanish sugar industry could have reaped enormous profits from Spain's neutrality in the war[3] had it not been for a succession of blunders, both on the part of the manufacturers themselves and the politicians responsible for the economy as a whole, who squandered every opportunity that came their way. During the final months of 1914 and the beginning of 1915, for example, the national sugar industry had succeeded in exporting almost 20,000 tonnes of sugar, mainly to Great Britain and Portugal, but the manufacturers, who had finally rid themselves of the economic millstone of all their surplus stocks, refused to believe that the conflict would last and thus during 1915 they reduced their contracts for sugar beet and sugar cane in a definitive attempt to re–establish the balance between supply and demand. The result was that at the end of the 1915 season the total sugar stocks throughout Spain were only 89,365 tonnes and the government feared that even the national demand could not be catered for.

Table 4.1. Spanish Sugar Production 1910–1935.

Year	Cane Tonnes	%	Beet Tonnes	%	Total Tonnes	%
1910	20300	22.2	71064	77.8	91366	100.0
1911	20294	19.1	86129	80.9	106424	100.0
1912	16175	10.4	138774	89.6	154949	100.0
1913	13231	8.2	148769	91.8	162000	100.0
1914	7376	5.0	140394	95.0	147770	100.0
1915	5595	5.2	101258	94.8	106853	100.0
1916	4264	3.7	111541	96.3	115806	100.0
1917	4583	3.7	119592	96.3	124176	100.0
1918	5712	4.4	123453	95.6	129166	100.0
1919	6278	5.1	117094	94.9	123372	100.0
1920	6760	4.5	142242	95.5	149102	100.0
1921	12033	7.0	159722	93.0	171755	100.0
1922	8097	5.9	130298	94.1	138395	100.0
1923	8454	4.9	164350	95.1	172804	100.0
1924	7661	3.4	216319	96.6	223975	100.0
1925	8704	4.2	200271	95.8	208975	100.0
1926	6718	4.2	250775	97.4	257474	100.0
1927	10551	4.5	221715	95.5	232267	100.0
1928	11610	4.8	230517	95.2	242217	100.0
1929	13561	5.7	223738	94.3	237300	100.0
1930	15756	5.9	249829	94.1	265285	100.0
1931	17912	5.7	315811	94.3	333724	100.0
1932	17459	5.9	276328	94.1	293787	100.0
1933	17431	7.1	226131	92.9	243562	100.0
1934	13097	4.2	296228	95.8	309325	100.0
1935	16831	5.8	274671	94.2	291502	100.0

Source: *Asociación General de Fabricantes de Azúcar de España*

From this moment on confusion reigned and the Exchequer excelled itself in promulgating a series of extempory measures and dictating a series of improvised norms which led to irreversible damage to the national economy as a whole, without doing anything to help the sugar industry take advantage of the international situation. To start with, a Royal Order

Table 4.2. *Spanish Sugar Imports and Exports 1910–1935 (tonnes).*

Year	Imports	Exports
1910	37	–
1911	42	271
1912	25	1
1913	37	10
1914	14	11472
1915	42	8755
1916	18330	2530
1917	39171	4581
1918	15194	3882
1919	29711	4311
1920	51644	661
1921	47263	1791
1922	37501	661
1923	736	7
1924	25299	61
1925	925	4
1926	414	20
1927	6444	235
1928	4835	28
1929	653	30
1930	138	77
1931	128	49
1932	3	16
1933	15	36
1934	35	35
1935	50	1

Source: *Memorias sobre el Estado de la Renta de Aduanas*

of 10 January 1916 opened wide the gates to the importation of foreign sugar, which for the rest of the war could enter the Spanish market on the same terms as the national product. If this were not sufficient, on 24 November 1916 another Royal Order, devised in order to forestall a rush of sugar away from the national market to the much more lucrative international one, forbade the exportation of sugar. With these measures the government not only frustrated the possibility of gaining large profits from increased exports, which the international situation promised and the potential excess in capacity of the national industry could have catered for, but also ensured that the home market was flooded by sugar from the Antilles, while Spanish manufacturers were cutting back on production for fear of finding themselves once more in a pre–1914 crisis situation if the price of sugar were suddenly to drop. In fact, faced with these measures, the sugar manufacturers reduced their contractual commitments for raw materials to the extent that from 1914 to 1919 national sugar production fell on average to something under 120,000 tonnes. This was far less than the 155,000 tonnes average achieved in the three years prior to the war and equally below the national consumption figures, which remained at around pre–war levels (Table 4.1). On the other hand, sugar imports, which had been virtually nil since 1900, resumed at around 25,000 tonnes during the war years 1916 to 1919 (Table 4.2). The resulting paradox was that a national industry that was potentially capable of overproducing could not even cater for its own home market at precisely a time when there was a serious shortage of sugar on the international market.

The overall loser in this absurd political charade was the Spanish economy as a whole because the sugar manufacturers were not slow to take advantage of the situation within the limits open to them. Between September 1915 and August 1918 the price of sugar rose from 85 ptas. per 100 kilos to 193 ptas. per 100 kilos, a rise substantially above that of the general price–rise index. Although the main beneficiaries of this rise were the sugar–beet growers, as can be seen in the price rises of cane and beet compared to refined sugar (Table 4.3), the manufacturers also profited from the situation, as the figures in Table 4.4 show, where the profits of the largest Spanish sugar firms are set out for three years. With these profits the companies were able to put their house in order after a financial crisis which had been deepening for several years, and even make substantial material reforms to their plants, thus being able to enter the post–war years on an even footing with their European competitors.

They did not, however, count on having to put up with several further arbitrary pieces of legislation on the part of the government. For instance, in line with an overall policy of price freezing, begun in the last months of 1916 with the intention of keeping down the cost of staple goods, a Royal Order of 24 September 1918 fixed a ceiling price for the sale of sugar. This law constituted an unprecedented and damaging governmental intervention as far as the sugar manufacturers were concerned, as they found that not only did the law apply retroactively to their stocks but that it did nothing to put a similar curb on the cost of many of the inputs necessary to their industry. In all, it formed one stumbling block more to taking advantage of the current international demand. The manufacturers, alarmed at growing state interventionism, drastically reduced their sugar beet commitments from the 66,000 ha. of the 1918–19 season to 41,000 for the 1919–20 season, at a time when the area of sugar cane planted had already been reduced to a minimum of 2,000 ha. since the beginning of the war. As a consequence the total 1919 sugar harvest amounted to no

Table 4.3. Prices of Cane, Beet and Sugar in Spain, 1913–1935.

	Cane ptas./ tonne	Beet Index	Sugar ptas./ tonne	General Index	ptas./ tonne	Index	Price Index
1913	34	100.0	40	100.0	73.3	100.0	100.0
1914	34	100.0	40	100.0	79.5	100.0	100.0
1915	34	100.0	40	100.0	90.6	124.0	119.0
1916	38	111.8	48	120.0	115.7	158.4	141.0
1917	45	132.3	65	162.5	117.9	161.4	166.0
1918	47	138.2	83	207.5	149.4	204.6	207.0
1919	54	158.8	90	225.0	163.6	224.0	204.0
1920	86	252.9	115	287.5	230.6	324.2	223.4
1921	58	170.6	78	195.0	143.0	195.8	184.6
1922	51	149.9	78	195.0	159.6	217.7	172.5
1923	54	158.8	95	237.5	163.5	223.0	170.9
1924	55	161.8	100	250.0	164.2	224.9	181.9
1925	55	161.8	88	220.0	154.7	217.3	185.0
1926	56	164.7	73	182.5	151.2	207.0	174.8
1927	51	149.9	70	175.0	146.4	200.5	167.9
1928	46	135.3	93	232.5	150.8	206.5	162.6
1929	54	158.8	103	257.5	152.4	208.7	167.7
1930	52	152.9	103	257.5	155.7	213.2	167.0
1931	68	200.0	90	225.0	149.5	204.8	168.8
1932	68	200.0	87	217.5	136.1	186.4	166.6
1933	47	138.2	77	192.5	148.4	203.2	159.1
1934	47	138.2	79	199.5	159.3	218.1	163.4
1935	48	141.2	80	200.0	159.8	218.9	164.2

Source: Compiled from *Memorias sobre el Estado de la Renta de Aduanas* and *Anuarios Estadisticos de España*.

Table 4.4. Profits of the most important Spanish sugar companies immediately before and during WWI (000 of ptas.).

	Sociedad General Azucarera	Azucarera del Ebro	Industrial Castellana
1912–13	59	40	1018
1913–14	3451	250	391
1914–15	4732	627	767
1915–16	16914	2487	1728
1916–17	18883	2416	2506
1917–18	–	2106	4383

Source: *Anuario Financiero y de Sociedades Anónimas de Espa.*

more than 117,000 tonnes, scarcely 70 per cent of the home demand, a demand which obviously had to be satisfied once again by increased importation.

Thus WWI came to end without the Spanish sugar industry having been able to reap the financial benefits which should have accrued to a neutral country in this international conflict. On the contrary, when the manufacturers' greatest problem was overproduction and many of their factories were being obliged to close down, and when the land–owners would have been only too ready to increase the areas of land given over to the cultivation of sugar beet if they were paid the price the international market would stand, they were obliged to sit back and watch sugar being freely imported from the Antilles.

Of the 45 sugar–beet processing plants and the 44 cane plants that existed in 1915 only 27 and 17, respectively, were working at the end of the war. Of the 100,000 and more hectares of sugar beet and 5,000 hectares of sugar cane that might have been cultivated, and indeed were a few years later, not even half that amount was grown during the war years. The only gain that the Spanish sugar manufacturers made during these years was a somewhat larger income from the higher prices caused by the war. What a poor profit from such an exploitable situation!

The dictatorship of Primo de Rivera: a controlled market

At the end of WWI the Spanish sugar industry was still having to put up with its traditional problems of potential excess capacity and high cost, but aggravated now by an ever–increasing policy of interventionism on the part of the government.

After the great efforts the industry had made to bring its technology up to date and to put its house in order financially, all made possible by the profits of the war, and after the arduous process of adapting output to the demands of the home market, the sugar manufacturers themselves could not now be held responsible for these high production costs. In fact, the exaggerated production costs of Spanish sugar were still due to the exorbitant prices of the raw materials, sugar cane and sugar beet, coupled with the high price and poor quality of the coal they were obliged to use. Thus, the principal inputs in the fabrication process accounted for more than three quarters of a typical Spanish sugar factory was very different indeed from its European counterpart (see Table 4.5).

The high prices of cane and beet were due in turn to the fact that they were cultivated in smallholdings, precluding the introduction of mechanized technology, added to which, the water needed to irrigate these plots was expensive. The 1906 Customs and Excise Act, passed in order to protect Spanish agriculture in general, had set up prohibitive tax barriers against the importation of sugar but had done nothing specifically to cut the costs of the farmer nor to encourage him to extend the cultivation of sugar beet to unirrigated land; this in fact was not to occur until many years later. As far as the coal used for the processing of the sugar was concerned, it came either from England or from Spanish mines in Asturias or in the Southwest of the peninsula. English coal was very expensive because of transport costs, particularly as the majority of the sugar plants were inland, while Spanish coal was very poor quality and transport was not that cheap, unless the factory happened to be in a mining area.

Table 4.5. Cost of Production of 100 kgs of sugar in a typical Spanish factory, 1922–24.

Item		Cost(ptas.)
Sugar Beet		103.60
Production Costs		32.40
Production Expenses	16.30	
Miscellaneous Exp.	3.30	
Quicklime	.50	
Coal	9.20	
Sacking	3.10	
Total Production Costs		136.00
Tax		45.00
Total Factory Cost		181.00
Sale Price		194.10
Profit		13.10

Source: *Archivos Azucarera San Isidro. Azucarera la Vega, Azucarera Antequerana.*

The greatest problem which the industry had to face at this time, however, was the interventionism of the government of the day, which, far from diminishing as the war ended, increased daily until the natural growth of the sector according to market norms was completely impeded. During the dictatorship of Primo de Rivera (1923–30) interventionism was the watchword in all the economic sectors. Not only were maximum prices imposed for many staple products, sugar included, but also a series of laws curtailed the possibility of free–market competition, resulting in dire inefficiency in the economic system as a whole, encouraging monopolistic tendencies and once more widening the gap between the Spanish economy and that of the rest of Europe.

The law which most affected the sugar industry was the Royal Order of 4 November 1926, by which the 'Comite Regulador de la Producción Industrial' was set up. This body was empowered to control the installation, extension, modification or geographical movement of all sugar factories according to criteria such as the production capacity of the sector in any area and consumer demand. A greater equilibrium between supply and demand was indeed achieved by these measures, but the behaviour of the 'Comite' resulted also in a curbing of technological progress and an imbalance in territorial distribution within the industry. Partly as a consequence of this law, the oligopolistic structure of the Spanish sugar industry, a structure which still endures today, came into being. Since its foundation in 1903 the General Sugar Company had been gradually losing its share of the market, until it could claim less than 40 per cent of the national production, but nevertheless it had always maintained its position as the leading single company in the industry. From 1923 onwards, however, the 'Ebro' group and the 'Industrias Agricolas' company, both of which had existed for some time before, became more and more powerful until, together with the General Sugar Company, they made up an oligopoly within the Spanish sugar industry.

Primo de Rivera's dictatorship was also responsible for the organization of corporate structures within the economy, which, in the sugar industry, resulted in the creation of 'Comisiones Arbitrales Mixtas' of sugar growers and manufacturers. The function of these commissions, at both local and regional level, was to arbitrate in disagreements and coordinate interests between the growers and the manufacturers. At first these commissions confined themselves to making sure that contracts undertaken between growers and manufacturers were fulfilled but later they extended their sphere of influence until they became the only channel for negotiation between all parties involved in the sugar sector. Henceforth the age–old conflicts between growers and manufacturers came to be settled within a much clearer framework than that which had previously existed.

Table 4.6. Spanish Sugar Factories 1900–1935.

	Factories	Cane Working Factories	Tons/Day	Factories	Beet Working Factories	Tons/Day
1900	30	23	155	48	40	156
1905	33	25	124	55	30	203
1910	36	29	65	49	34	288
1915	44	17	38	39	27	238
1920	24	20	36	40	31	228
1925	25	17	64	43	35	598
1930	27	15	125	47	38	491
1935	29	14	136	54	41	590

Note: Average for all factories based on a season of 100 days.
Source: *Memorias sobre el Estado de la Renta de Aduan.*

Despite the price which the industry had to pay, the basic objectives of helping the national sugar industry and meeting the demands of the home market were fully achieved. The 1931 season saw a record production of 333,724 tonnes, a figure that was not repeated for another 20 years, and the per capita consumption of sugar stood at around 12 kilos, almost double that of just before the war.

As far as the growers were concerned, great improvements were made in the cultivation of sugar cane, thanks mainly to the introduction of new strains, although little progress was made with sugar beet. Despite this, however, the growers were able to benefit from the experience gained in having to apply new technology to their traditional farming methods. The factories found themselves able to fulfil their production capacity, increasing their output to 600 tonnes per day in the beet sector and 125 tonnes per day with cane, although some of these factories were little more than mills or *trapiches* producing treacles and raw sugar (see Table 4.6.).

Table 4.7. Spanish Sugar Beet. Agricultural and Industrial Yields.

	Beet/Ha (Tonnes)	Sugar Yield	Sugar/Ha (Tonnes)	Cost of beet per 100 Kgs of sugar (Ptas)
1915–16	21	12.73	2.67	31.42
1916–17	19	12.56	2.39	38.22
1917–18	21	11.50	2.42	56.52
1918–19	17	13.66	2.32	60.76
1919–20	17	11.71	1.99	76.86
1920–21	25	11.56	2.89	99.48
1921–22	18	10.15	1.83	76.85
1922–23	20	13.40	2.68	58.20
1923–24	20	11.99	2.41	79.23
1924–25	18	12.26	2.21	81.60
1925–26	17	12.28	2.09	71.66
1926–27	20	11.52	2.30	63.37
1927–28	20	12.24	2.46	57.19
1928–29	20	10.65	2.13	87.32
1929–30	20	11.90	2.38	86.55
1930–31	21	12.24	2.57	84.15
1931–32	20	12.60	2.52	71.43
1932–33	17	11.84	2.01	73.48
1933–34	18	11.11	2.00	69.31
1934–35	22	12.97	2.85	60.91

Source: Compiled from *Memorias sobre el Estado de la Renta de Aduanas*.

The sugar industry during the 2nd Republic

The worldwide monetary crisis of the 1930s hardly made itself felt in Spain, although the fall of the dictator Primo de Rivera brought with it a radical change in economic policy. Public spending, which had been the principal growth mechanism of the dictatorship, was drastically curbed, while some salaries that had been virtually frozen for 10 years were substantially increased. But all this merely restructured the economy in favour of the consumer industries while doing nothing to decrease the growth rate of the economy as a whole.[4] Policy towards the sugar industry in particular showed no change and this was reflected in the overall response of the industry. The ample 60 ptas. per 100 kilos margin of protection, which had been established by the Dictatorship in 1928, remained in force and, despite protests from both growers and manufacturers, sugar continued to be heavily taxed at 45 ptas./100 Kilos, in fact, sugar revenue accounted for almost 5 per cent of the national income at this time. On this basis the industry maintained a high production level, varying according to the home market, and governed by contingency agreements between the most powerful manufacturing groups. Due to the protectionist trade barriers sugar prices remained way above the general price–index, as had occurred in years before. Once again, the prime movers of this were the growers rather than the manufacturers. The usual conflict of interest between the cultivation of cane and that of beet was held in abeyance during this period as the strains of cane introduced into Spain from Java in 1925 turned out to be sufficiently equal to the subtropical areas of Spain as to occupy almost the whole of the potential geographical limits of cultivation in the peninsula, a quantity not seen since the beginning of the century, when sugar beet was not yet being cultivated extensively on irrigated land.

After ten such years of moderate stability, during which both growers and manufacturers enjoyed a relatively harmonious co–existence, the final step was taken to control the annual output of sugar. An Act of Parliament on 23 November 1935 gave powers to the 'Comision Mixta Arbitral' to regulate the quantities of sugar beet and sugar cane that might be cultivated in any area of the peninsula and also to determine the price which these raw materials could command, according to a scale of pulp density. The 'Comision' was also to draw up the general terms of contracts of sale. Once more the setting up of new factories and the expansion or removal of already existing ones was forbidden. It was furthermore forbidden to close down any plant already in production. Guarantees were also provided to meet the demands of the home market at any time and a commission was set up to study the possibility of bringing down the price of sugar, while at the same time increasing production by stimulating public demand.

On the whole, this policy, which anticipated to a great extent the modern laws governing the annual sugar seasons, can only be criticized on one major count, that of increasing isolationism from the rest of Europe. While it is a fact that the greater number of sugar–producing nations within and outside Europe were doing the same thing and that most of the international markets were protected and would have remained closed to Spain in any case[5], it still remains true that the guarantied 60 ptas. per 100 kilos protection (40 per cent of the wholesale value) was too high to persuade the industry as a whole that it was vital to increase productivity.

Table 4.8. Sugar Cane, Agricultural and Industrial Yields 1915–1935.

Year	Cane/Ha Tonnes	Sugar Yield %[1]	Sugar/Ha Tonnes[1]	Cost of Cane per 100 kgs of Sugar (ptas).
1915	38	8.76	3.33	38.81
1916	37	9.61	3.56	39.54
1917	34	7.18	2.44	62.67
1918	45	6.61	2.97	71.10
1919	35	9.71	3.40	55.61
1920	39	9.28	3.62	92.67
1921	55	9.36	5.15	61.97
1922	43	8.69	3.74	58.69
1923	40	9.13	3.65	59.15
1924	42	7.10	2.98	77.46
1925	44	8.04	3.54	68.41
1926	36	8.46	3.05	66.19
1927	41	8.20	3.36	62.20
1928	40	8.30	3.32	55.42
1929	64	8.05	5.15	67.08
1930	70	8.40	5.88	61.91
1931	60	8.57	5.14	79.35
1932	61	8.40	5.12	80.95
1933	65	8.52	5.54	55.16
1934	63	7.71	4.86	60.95
1935	65	8.08	5.85	59.40

Note: [1] White Sugar.
Source: *Memorias sobre el Estado de la Renta de Aduanas.*

The quantity of sugar extracted from the sugar beet was coming closer to that of the rest of Europe, but the great problem continued to be in the yield per hectare, which in Spain was only around 20 tonnes per ha., with high production costs to boot. Thus from 1930 to 1934 the average refined sugar yield per hectare did not exceed 2.40 tonnes, a figure only slightly more than half the European average. As far as sugar cane was concerned the deficit was even greater when compared to countries such as Java or Hawaii (Tables 4.7 and 4.8).

Nevertheless, despite all its false starts, and the problems it had to overcome, by the outbreak of the Spanish Civil War in 1936, the sugar industry had achieved an organization and infrastructure which few other sectors of the economy had acquired by that time.

References

1. M. Martín, *Azúcar y Descolonización,* Universidad de Granada, Granada, 1982.

2. J. Ceballos Teresí, *El problema azucarero,* Madrid, 1914.

3. The favourable effects of the First World War on the Spanish economy have been studied in the excellent work by S. Roldán and J.L. García Delgado, *La formación de la sociedad capitalista en España,* 1914–1920, 2 vols, Madrid, 1973.

4. J. Palafox, 'La gran depresión de los años treinta y la crisis industrial española', *Investigaciones Económicas,* Madrid, enero–abril 1980, 5–46.

5. B.C. Swerling, *International Control of Sugar, 1918–41,* Stanford, 1949 and V. Gutierrez Valladon, *El problema mundial del azúcar,* Madrid, 1936.

5

Growth and Crisis of the Brazilian Sugar Industry, 1914–39

Tamás Szmrecsányi

During the inter–war period the Brazilian sugar industry experienced a strong productive expansion, some important structural changes and a considerable crisis of overproduction. Despite their intensity, none of these processes had any significant consequences either on the country's position in the international sugar market, or on the level of the country's external revenues. This was so because, unlike her situation in former centuries, Brazil had by this period become a marginal exporter of sugar, and because the volume exported accounted only for a small part of her overall production. Her sugar production was mainly, or almost completely directed towards the internal market.

The data of Table 5.1 show that production increased substantially, and almost continuously, after the beginning of WWI. This trend was only initially, and partially, induced by the growth of exports, which, only had a major effect on the increase of production from 1914 to 1923. After this the relevance of exports declined considerably, becoming through time an ever less important consideration. Except during the war and immediate post–war years the main determinant of the observed increase of Brazilian sugar production has always been the expansion of the country's domestic consumption.

The development of an internal market in Brazil was, at the time, a relatively recent process. It had begun some decades before, in the second half of the nineteenth century, and it was basically due to the diffusion and differentiation, and to the productive growth, of the coffee export complex. Between 1850 and 1930, this remained the leading sector of the country's emerging national economy. Its expansion and diversification gave rise over time to both a progressively more important manufacturing sector and to the establishment, within Brazil, of a new regional division of labour.

The Brazilian sugar industry

Table 5.1. Brazilian Sugar Production and Exports, 1910–40.

Crop–Years	Output (*)	Index Numbers	Civil Years	Exports (*)	Indxe Numbers	% Output Exports
1910/11	5529	93	1911	603	114	10.9
1911/12	5039	84	1912	80	15	1.6
1912/13	5557	93	1913	39	7	0.7
1913/14	5965	100	1914	531	100	8.9
1914/15	6618	111	1915	986	186	14.9
1915/16	6672	112	1916	907	171	13.6
1916/17	7566	127	1917	2303	434	30.4
1917/18	8025	135	1918	1927	363	24.0
1918/19	8608	144	1919	1707	321	19.8
1919/20	11588	194	1920	1819	343	15.7
1920/21	12128	203	1921	2868	540	23.6
1921/22	14341	240	1922	4202	791	29.3
1922/23	14209	238	1923	2553	481	18.0
1923/24	14372	241	1924	574	108	4.0
1924/25	15370	258	1925	53	10	0.3
1925/26	12489	209	1926	286	54	2.3
1926/27	15592	261	1927	308	58	2.0
1927/28	13869	232	1928	501	94	3.6
1928/29	15700	263	1929	248	47	1.6
1929/30	19601	329	1930	1408	265	7.2
1930/31	16996	285	1931	185	35	1.1
1931/32	17125	287	1932	674	127	3.9
1932/33	16270	273	1933	424	80	2.6
1933/34	16602	278	1934	398	75	2.4
1934/35	16555	278	1935	1448	273	8.7
1935/36	17900	300	1936	1380	260	7.7
1936/37	14997	251	1937	5	1	0.0
1937/38	16743	281	1938	135	25	0.8
1938/39	18340	307	1939	806	152	4.4
1939/40	19632	329	1940	1102	208	5.6

Note: (*) Production and Exports in 1000 bags of 60 kgs.
Source: T.Szmrecsányi, O Planejamento da Agroindústria Canavieira do Brasil (1930–75), São Paulo, 1979 , 504.

Through a combination of all these processes, the core of the Brazilian economy was transferred from the Northeast, where it had rested during most of the colonial period as well as the first decades of the country's political emancipation, to the Centre–South, and more specifically to the state of São Paulo.[1] By the same token, sugar, which had been for centuries Brazil's most important export staple[2], became virtually overnight a secondary product. Its relevance remained unchanged only at the regional level, where it tended to become more important over time. And this was so in the very same Northeast, more particularly in the state of Pernambuco, which retained for several other decades the absolute leadership of Brazilian sugar production.

By the end of the 19th century the sugar mills of this region ceased almost completely to sell their production in the foreign markets. Instead Pernambuco's sugar went to the Centre–South of the country particularly to Rio de Janeiro and São Paulo. This change was only partially and temporarily reversed during the years 1914 to 1923. This new market orientation had clear benefits given the strong competition and low prices which characterized the international market in that period. Although smaller and less dynamic, the internal markets of the Centre–South presented the advantage of being sheltered from external competitors and, therefore, were, at least in principle, capable of providing higher profits for Brazilian sugar producers[3]

However, these higher profits, coupled with the constant expansion of domestic demand, attracted new internal competitors to the sugar industry. This attraction was further stimulated by the recurrent and ever growing overproduction crises in the coffee export economy. The new producers were mainly from the states of Rio de Janeiro and São Paulo and therefore, had the advantage of being closer to the main consuming markets than the sugar–mill owners of the Northeast. The increasing output of these states would become one of the main factors in sparking the overproduction crisis which erupted in the Brazilian sugar industry at the end of the 1920s.

The origins and evolution of this crisis, its basic features and main consequences are the subject of this paper. However, before considering these issues, it is important to note that the highest level of production before the Second Word War was attained during the crop–year 1929/30. This points to the relative success of the Brazilian government's intervention in the sugar industry during the 1930s. It is also must be stressed that this intervention was only rarely based on export of the production surpluses, as happened, for instance, in the crop–years 1934/35 and 1935/36. Almost from the beginning, its main instruments were, on the one hand, the stimulation of large–scale alcohol production and, on the other, the establishment and maintenance of a comprehensive and rigorous system of sugar and cane production quotas, and of administered prices for both. It was through the utilization of these two complementary policy instruments that the Brazilian government succeeded in saving the greatest part of the country's sugar industry from otherwise unavoidable bankruptcy, thus preparing the way for its future growth and consolidation.

Although it only became dramatically apparent at the end of the 1920s, the sugar industry's overproduction crisis was already 'in the air' from the turn of the century, when its exports were virtually eliminated from the international market.[4] This was due in great part to the industry's obsolete and high–cost production methods, which did not allow it to compete with the protected and subsidized beet sugar producers on the one hand, and with technically more advanced and economically more efficient cane sugar producers, like Cuba, on the other. For producers from the Northeast difficulties were compounded by the fact that increasingly they were having to compete in the domestic market with the larger, better equipped and more advantageously located producers of the Centre–South. Things did not become worse earlier for the Northeastern sugar–mill owners thanks to WWI which reopened temporarily the possibility of exporting the main part of their production. These opportunities, however, not only induced an increase of overall production, but they also resulted in the expansion of productive capacity. The latter would prove to be a burden when it became apparent that the export opportunities had been only temporary.

Inter–regional competition for the markets of the Centre–South became fiercer through time. Northeastern producers were not, however, immediately displaced due to the phytosanitary problems which plagued the São Paulo and Rio de Janeiro sugarcane growers throughout most of the 1920s. Nevertheless, at the end of that decade, the definitive overcoming of these problems coincided, not only with all–time record levels of sugar production in the Northeast, but also with the almost simultaneous eruption of the most serious overproduction crisis to ever hit the coffee sector together with the Crash of 1929 and the onset of the Great Depression.

These two crises had an immediate impact on effective demand in general, and on sugar consumption in particular. Sugar prices on the Rio de Janeiro commodity markets plunged by almost 65 per cent between March and October 1929. Net prices (less transport costs) obtained by producers costs fell below the average level production costs in the Northeast, instantaneously putting out of business most of the sugar manufacturers and cane growers of that region.

The new low prices were, however, still remunerative for the producers of the Centre–South, in particular for those closer to the cities Rio de Janeiro and São Paulo. In the state of São Paulo there was moreover a further incentive for the expansion of local sugar industry, because of the overproduction crisis faced by coffee growers. Unprofitable coffee lands and idle capital could be, and were, rapidly converted respectively into cane growing and new and/or larger sugar mills. These changes occurred within a very short time. For example, the number of *usinas* in São Paulo, which had not changed between 1910 and 1920, increased from twelve to twenty in the 1920s and by another sixteen during the depressed 1930s. Similar trends could be observed with regard to the overall sugar production of the state of São Paulo, which grew by 41 per cent in the 1920s and by 111 per cent in the 1930s.

The days of the Northeastern sugar industry seemed to be numbered. Something had to be done in order to avoid its partial bankruptcy. This effectively happened thanks to the 1930 political revolution, which toppled from power the oligarchical regime of the First Republic, dominated by the agrarian, commercial and financial interests related to the coffee export economy.[5] The new government put into power by the revolution promptly adopted a set of interventionist policies, the main intention of which was to warrant the survival, at the very least, of the Northeastern sugar industry.

With some important changes, these policies remained effective until the end of the 1960s. They were responsible not only for the survival of the Northeastern sugar industry but also for the further overall growth and diversification of that industry in the country as a whole, more particularly in the Centre–South. An important factor, which seems to be forgotten nowadays by their main beneficiaries, is that those interventionist policies were expressly asked for by the producers themselves, mainly, but not only, those of the Northeast.

Such policies included, in the first place, measures for fostering the production and the consumption of alcohol (ethanol) as a fuel for automotive vehicles, in order to permit a productive use for the huge and increasing surpluses of sugarcane. In the second place, and perhaps even more importantly for those (and for any other) times, they led to the establishment, in 1935, of a comprehensive and rigorous system of production quotas and administered prices both for sugar and cane. These sets of complementary measures contributed in turn to transform the Brazilian sugar industry and its various interrelated parts into something like a centrally planned and centrally managed sector of the country's national economy.

It is not possible here to discuss in detailed the genesis of these policies.[6] Suffice it to say that those policies were the most rational, and probably the only ones possible, at a time when there was a great excess capacity in both the agricultural and the manufacturing sectors of the industry and few possibilities of exporting surpluses at minimally acceptable prices. Alcohol had already been produced in Brazil for decades, as it was (and still is) an almost necessary by–product of cane sugar manufacturing, obtained through the further processing of the residual molasses. At this time, however, that production was still incipient because of a lack of markets for alcohol. Except for some fuel consumption by the trucking fleets of a few larger sugar mills, sugarcane alcohol produced in Brazil was consumed in minimal quantities as an input by the emerging chemical, pharmaceutical and beverage industries. The government's intention at the beginning of the 1930s was to transform it into a mass–consumption good by having it used as an additive to automotive gasoline, all of which was imported in growing quantities.

This measure, if implemented, would therefore produce at least two very important effects. On the industry's level, it would provide an alternative use for sugarcane, by upgrading alcohol from a byproduct of sugar to cane's main derived product, giving the industry a greater capability to face the less favourable market conditions for sugar. Secondly, in macroeconomic terms, it would allow the country to save scarce foreign exchange, through a reduction of its gasoline (and later petroleum) imports by 5 per cent, 10 per cent or even more.

In the late 1930s, and up to the beginning of WW II, this was to be one of the most successful policies, absorbing part of the sugar industry's excess capacity without the need to produce more sugar, and at the same time helping to balance the country's external accounts. Brazil's alcohol production grew by almost 134 per cent during the 1930s, from a yearly output of 40 million litres between 1930 and 1935 to 94 million litres in the crop–year 1939/40. It did not increase faster in these years because of a lack of capital (and maybe also of interest) in the private sector, the scarcity of local–made capital equipment for alcohol distilleries[7] and the difficulties of importing such equipment, due to a lack of foreign exchange.

A similar judgment can be made with regard to the establishment of governmental control on the production and on the internal prices of sugar, alcohol and cane. This extreme policy was adopted by the Brazilian government only after other partial measures had been tried without success. For example, there was an attempt to create buffer stocks which would later be either exported or sold on the internal market. There was also a project to levy a tax on surplus production and on production in general. Neither measure had the desired effect. The decision to limit sugar production to no more than an average of previous five years was intended not only to eliminate gluts, but also to make available more raw material to be transformed directly into alcohol.[8] That decision could, however, only be implemented if and when production quotas for sugar were established and controlled at mill level by a central authority.

The first important step in this direction was taken in 1933 with the creation of the *Instituto do Açúcar e do 'Alcool*, which would be better known by its initials IAA. This agency of the federal government still exists today, although having been terribly disfigured (as everything else in Brazil) by the twenty–one years of military dictatorship (1964–1985), and by the now more than ten years adventurous alcohol program called PROALCOOL. A second, but an even more decisive, step was the establishment in the 1935 of the sugar production quotas themselves. This measure was preceded and complemented by the legal prohibition of installing any mill, or of expanding existing ones without the previous and explicit consent of the Institute for Sugar and Alcohol. Transgressors of this precept were subject to pecuniary fines and to confiscation of the new equipment. All existing equipment and installations had to be registered with the state agency. This together with the average of the previous five years' current production was the basis upon which the agency fixed each sugar mill's quota. The overall limits of production by state (and, consequently, by mill) were to be fixed annually through the so–called crop–plans, gauged according to the prospects for domestic demand and to the level of existing sugar stocks within the country.

All these measures were rigorously applied in the 1930s, contributing decisively to the stabilization of Brazilian sugar production (see Table 5.1), by equating supply to domestic demand, and by establishing prices which were remunerative for the industry and attractive for the final consumers. Within a few years, state intervention in the sugar industry had become the norm, at least at the level of the *usinas*. With time, even the tens of thousand tiny, simple *engenhos* or *banguês* were also put under the administrative control of IAA.

A problem which still remained to be solved (and which would only be solved definitely in the 1940s) was that of the relationship between the *usinas* and their sugarcane *fornecedores* , the suppliers of part of their cane needs. Specially in the Northeast, many of these cane suppliers were former *senhores de engenho* (owners of the smaller, semi–artisanal mills) which had abandoned sugar manufacturing to specialize in cane cultivation. These producers, who collectively had some political influence, did not want to accept a subordinate position *vis–à–vis* the larger mills, which, by having their own sugarcane plantations and by becoming evermore self–sufficient in raw material, were able to dictate prices and other cane supply conditions. Relations between *usineiros* (greater–mill owners) and their *fornecedores* reached a dangerous level of antagonism by the end of the 1930s. This in turn forced the Institute to intervene by establishing a system of cane supply quotas as well as minimal prices to be paid for purchased sugarcane. By this system, implemented in the 1940s, the *fornecedores* became entitled to supply one half, or even more, of the cane to be ground at the *usinas*. This progressive economic and social legislation would be practically overturned at the beginning of the 1970s.

Such were the main effects and the more immediate results of governmental intervention in the Brazilian sugar industry. At the same time that it helped this industry to overcome a grave overproduction crisis, and to solve several institutional problems, this intervention also left other, equally important problems, completely untouched, aggravating some of them, and even gave rise to entirely new ones. By attempting to salvage and to consolidate the status quo, it had the effect of hindering, or postponing, structural changes in the Brazilian sugar industry, which otherwise would have been arrived at faster and/or easier.

The production limits by states and the production quotas by mills were decisive in permitting, not only the survival, but also the continuing supremacy of the Northeastern sector of the Brazilian sugar industry. Thanks to governmental intervention, this region benefited from privileged access to the rich domestic markets of the Centre–South. For this reason, the producers of that latter region were not allowed to expand their output and to increase their share of these markets according to their economic potential, which, at that time, was more favourable than that of the Northeastern sugar mills. Such an 'abnormal' situation could only prevail (and be imposed) at relatively 'normal' times. Whenever these conditions changed, the actual comparative advantages would surface with great force. This really happened in the 1940s, with the eruption of WW II, the effect of which was greatly to increase the participation of the Centre–South in total Brazilian sugar production. By the next decade it had outstripped the Northeast.

Another less satisfactory consequence of governmental intervention in the Brazilian sugar industry was a certain slowing down in the rate of technical progress during the 1930s and later. The allotment of production quotas became an impediment to the increase of individual production levels, and to the consequent realization of economies of scale. On the other hand, the fixing of centrally administered prices, by providing a minimal return even to the least efficient sugar production units, as well as a surplus profit to the more efficient ones, acted as a disincentive to obtaining higher productivity and/or of lower cost levels both in the agricultural production and in the industrial processing of sugarcane.

These problems, to be sure, were not new in Brazil. Their origins indeed went back to the failure, in the second half of nineteenth century, of the attempts at substituting the small–scale and technologically archaic *engenhos* by central milling, as had already occurred in other cane sugar producing areas of the world.[9] Such attempts did not succeed in Brazil for various reasons, one of the main of which undoubtedly was the unwillingness of the *senhores de engenho* to become mere cane suppliers to central mills which they neither owned nor controlled. The main result of this resistance, supported by the new regional governments which took power with the advent of the Republican regime in 1889, was the emergence and diffusion of the *usinas*. These production units, however, proved to be little more than larger size and technologically more modern *engenhos*. Their owners, like the *senhores de engenho* of the past, continued to be both cane and sugar producers, which meant that there was no division of labour between the agricultural raw material production on one side and the industrial processing of the final product on the other. Capital and other productive resources had to be shared between both, and this undoubtedly represented a limitation to the process of industrial and/or agrarian capital centralization.

Table 5.2. Comparison of the Production Levels of the Ten Largest Brazilian and Cuban Sugar Mills in the early 1930s.

Countries & Enterprises	Milling Capacity (Tons/24h)	Production (1000 bags of 60 kg)				
		1928/29	1929/30	1930/31	1931/32	1932/33
(1) CUBA						
1 Jaronu	12500	2000.4	1855.8	1190.2	1140.8	845.0
2 Manati	10000	1836.6	1512.5	1056.3	1007.6	627.1
3 Delicias	8480	2030.2	2001.6	1240.0	1137.6	763.7
4 Moron	8370	1895.0	1232.6	1331.2	878.5	874.5
5 Preston	8370	2489.1	1588.2	1286.0	1034.1	727.1
6 Boston	8000	1930.7	1434.8	1061.5	806.1	588.3
7 Hershey	7500	1248.5	1262.5	1016.4	1268.8	1004.0
8 Stewart	6700	1472.1	1319.5	1104.2	809.0	742.3
9 Chaparra	6700	1502.6	1302.6	916.4	864.5	643.4
10 San German	6100	1332.4	1194.1	884.0	662.8	513.7
(2) BRAZIL						
1 Catende	1768	248.1	442.6	225.6	400.2	295.1
2 Tiuma	1687	253.7	270.3	217.9	219.1	191.1
3 Central Leão	1466	231.1	400.7	282.8	235.8	253.9
4 Barreiros	1460	109.2	75.5	78.4	121.8	114.
5 Brasileiro	1429	60.5	138.4	110.7	91.5	102.0
6 União e Ind	1300	158.4	165.4	134.5	156.5	119.5
7 Junqueira	1300	67.0	115.1	106.3	164.7	142.8
8 S Terezinha	1600	60.0	128.0	84.0	190.0	157.1
9 Serra Grande	1247	177.3	322.2	176.0	188.2	247.7
10 S J da Varzea	1210	94.4	130.0	53.6	54.4	37.2

Source: IAA, *Anuário Açucareiro para 1935*.

Although the Brazilian *usineiros* became very big landowners and, through this politically and socially very powerful figures, they were up to the end of the 1930s, owners of relatively small sugar mills. This can be seen in Table 5.2, which presents comparative data of the installed capacity and production levels of the ten greatest sugar processing units of Brazil and Cuba. The figures speak for themselves, especially if we take into account of the fact that the declining production levels of the Cuban *centrales* in the early 1930s were the result of a deliberate policy of restraining output in view of the international overproduction crisis. Despite this, the sugar production of the Cuban mills remained several times greater than that of their Brazilian counterparts. Relatively small scale manufacturing, together with low productivity levels of cane farming in Brazil, help to explain why the country's sugar industry was not competitive on the international market, a situation that still prevails[10]

Due to the high capital requirements of large–scale sugar and/or alcohol manufacturing, investment in these activities seems far from compatible with great agricultural land acquisitions. While these may, or perhaps even should,[11] be decentralized, investment in sugar and/or alcohol manufacturing has to be further concentrated in ever larger and technologically more efficient plants or production units. From this we may conclude that the prevailing vertical integration of the Brazilian sugar and/or alcohol industry — that is, the bringing together within the same enterprises of both manufacturing and agricultural production, does not necessarily represent the best and cheapest way of developing either of them.

In order to increase productivity and efficiency of both the agricultural and the manufacturing segments of Brazilian sugar and/or alcohol industry, perhaps the most urgent tasks continue to be the establishment and the deepening of a new division of labour between both. A division of labour by which they would become capable, each one in its particular framework, not only of a greater specialization and of a more voluminous production, but also of a further diversification. This is the process which, according to the able demonstration of Edith Penrose, always goes together with capitalization and further growth.[12].

It is always worthwhile to recall in this regard that many other products can be obtained from land beside sugarcane, and that sugar and/or alcohol are not the only two commodities which can be extracted from this raw material. And finally, we may mention the fact that, in contemporary capitalism, large–scale but relatively unproductive monoculture has as little economic and social virtues as small–scale and relatively high–cost manufacturing. These are conditions which were crystallized in the 1930s, and which continue to prevail in the Brazilian sugar and/or alcohol industry of today. How to overcome this situation is a challenge to be faced now or in the very near future.

References

1. The best account of the Paulista coffee economy's expansion is Pierre Monbeig, *Pionniers et Planteurs de São Paulo,* Paris, 1952. For its impact on industrialization and regional concentration see, Sergio S. Silva, *Expansão Cafeeira e Origens da Indústria no Brasil,* São Paulo, 1976 and Wilson Cano, *Raízes da Concentração Industrial em São Paulo,* 2nd ed., São Paulo, 1983.

2. Luis Amaral, *Historia da Agricultura Brasileira*, vol.II, São Paulo, 1940, 61 and ff.

3. These points have been first stressed by P.I. Singer in the chapter on Recife of his *Desenvolvimento Economico e Evolução Urbana*, São Paulo, 1968, 313–314.

4. Most of this and of the following paraghaphs are based on my book, *Planejamento da Agroindústria Canavieira do Brasil (1930–1975)*, São Paulo, 1979, 161 and ff.

5. There is a vast and ever growing literature on the nature and consequences of that revolution. The best approaches to the subject, in my opinion, continue to be those of Boris Fausto, *A Revolução de 1930 – Historiografia e História*, 5th ed., São Paulo, 1978; Luciano Martins, *Pouvoir et développement économique – formation et évolution des structures politiques au Brésil*, Paris, 1976; and Eli Diniz, *Expansão, Estado e Capitalismo no Brasil: 1930–1945*, Rio de Janeiro, 1978.

6. See my *Planejamento da Agroindústria*. Also Oriovaldo Queda, 'A Intervenção do Estado e a Agroindústria Açucareira Paulista', unpublished PhD thesis, University of São Paulo, 1972.

7. This equipment was partly being manufactured in Brazil from the 1920s onwards. Barjas Negri, 'Um Estudo de Caso da Indústria Nacional de Equipamentos: análise do grupo Dedini (1920–1975)', unpublished M.A. dissertation, Universidade Estadual de Campinas, 1977.

8. For those who are not sufficiently familiar with cane/sugar processing it may be useful to observe that alcohol can be obtained not only as by–product derived from molasses – the so–called residual alcohol – but also directly from cane–juice, which therefore does not need to be transformed into sugar and molasses. On the other hand, it is always important to distinguish two kinds of alcohol or ethanol: a less elaborated one, or hydrated alcohol, traditionally used as an industrial raw–material or input, and nowadays in Brazil as a (rather poor) substitute for gasoline, and a further processed one, or anhydrous alcohol (over 99.50 GL), used for decades in Brazil, and now also in the USA, as an oxygenating additive to gasoline. It was the production and consumption of the latter, as well as direct alcohol production, that the Brazilian goverment of the 1930s intended to promote. This policy remained unchanged up to the end of the 1970s, when, at the instance of sugar mill owners of São Paulo, and with the support of the automobile industry, the military government decided to promote the mass production and consumption of hydrated alcohol. While Brazil continued to depend on large petroleum imports and while the international prices of this product remained high, everything went well. But nowadays the country has to cope with a huge overproduction of both alcohol and gasoline, not to speak of that of sugarcane, a problem which it was so difficult to erradicate in the 1930s.

9. In this regard, see, among others, the studies on the Caribbean (1840–1905) by Christian Schakenbourg, on South Africa by Peter Richardson, and on Australia (1862–1906) by Adrian Graves, which figure in B. Albert and A. Graves (eds.), *Crisis and Change in the International Sugar Economy, 1860–1914*, Norwich and Edinburgh, 1984, 88–89, 244–46, 256, and 276–79.

10. A similar phenomenon is occurring presently with the policy of substituting hydrated alcohol for gasoline, a policy which only became possible through heavy governmental subsidies for producers and consumers. At the present cost and productivity levels, alcohol is very far from being competitive with gasoline.

11. One who thinks so with good reasons is José Gomes da Silva, *A Agroindústria Canavieira em Países Selecionados:sistemas de produção de pequenos e medios agricultores*, Campinas, 1979.

12. E.T. Penrose, *The Theory of the Growth of the Firm*, 2nd ed., Oxford, 1980 , especially chapters V–VII.

6

The Peruvian Sugar Industry 1918–1939: Response to the World Crisis

Bill Albert

The Peruvian sugar industry's response to the crisis of the interwar period was varied. There were strenuous, and on the whole, abortive efforts to secure preferential export markets. Attempts to elicit material support from the state also failed. Faced with a massive fall in prices and a restricted number of possible markets, local producers, who in better times had rarely competed directly against one another, began to battle fiercely over both domestic and export markets in neighbouring Bolivia and Chile. This led within a short time to many smaller estates being forced out of production and the increased domination of a few large producers in the northern valleys, a trend which had been discernible from the late nineteenth century. But it was not size, location or even strong foreign links which made for survival. As will be shown, all these factors were important, but the key to the industry's survival and its restructuring was the phenomenal increase in labour productivity coupled with the substantially lower wages paid by the northern growers than their fellow planters in the central valleys. For all its impressive technological sophistication, the success of the sugar industry depended in the end on the planters' ability to control labour and maintain low wages. This did not, however, take place without resistance from the workers, resistance which became widespread and for the planters menacing in the early 1930s. It was at this point that the state came to the aid of the industry. Its repression of the workers and their organizations provided the necessary compliment to the planters' tactic of shifting the cost of lower sugar prices onto the workforce.

Access to export markets

The principal cooperative efforts of the country's sugar growers in these years were on two main fronts; trying to ensure and improve their position in foreign markets and putting pressure on the government for direct and indirect assistance. In neither were they very successful, despite the considerable political clout of the landed oligarchy, of whom the sugar growers were a powerful element. The lack of success was more understandable with respect to export markets, because while the state was generally sympathetic and gave sugar growers diplomatic backing, Peru did not have any political leverage internationally. Concern centred on the Chilean and British markets, especially as the United States, which had been of major importance for over a decade, was from 1925 effectively cut off from Peruvian sugar by restrictive commercial policy.[1]

Chile had been a major market since the colonial era, and its proximity seemed to ensure its continued dependence on Peruvian sugar. However, growers began to be anxious about this 'safe' market in the late 1920s when the Chilean government started to promote beet sugar production.[2] It was the calamitous impact of the 1929 Crash on the Chilean economy which most seriously alarmed Peruvian exporters. Chile's two principal commodities were badly hit and the total value of her exports fell by 80 per cent.[3] The level of imports had to be drastically cut, inconvertibility was declared, and wide ranging exchange controls imposed. Worried that the Chileans would restrict sugar imports, Peruvian producers pushed for a trade treaty. It was signed in 1934, but proved of little value as Chile was unable to find cheaper sugar elsewhere, and Peru continued to supply virtually all her needs. However, that the treaty was agreed to at all against strong nationalist objections in Peru points to the improved political position of the sugar interests who had not fared very well under the Leguía regime (1919–1930).[4]

Peru's other major outlet was the large British market, which had adhered to free trade since the mid–nineteenth century. This had been of great importance, for unlike many cane sugar exporters, nowhere did Peru enjoy preferential access for her sugar. In the post–war years Peruvian producers had cause for concern for this market began to be closed off by tariffs, colonial preference, and subsidies to a new local beet sugar industry.[5] After protracted talks an agreement was finally signed in 1936, but little more was offered to Peru than most favoured nation status, while Britain obtained a number of valuable tariff concessions.[6] However, as Peru received a generous allocation under the 1937 International Sugar Agreement and in the same year a much improved quota in the US market, the British treaty, like that signed with Chile, was not all that important.[7] Once again, however, the government had shown itself willing to champion the cause of the sugar producers, offering tangible concessions in the process. The record of the state's domestic support was not as clear cut.

Domestic policy

On the national level sugar producers through the *Sociedad Nacional Agraria* (SNA) fought for a wide variety of concessions and reforms, mainly over taxation. However, during the 1920s Augusto Leguía, was not only unresponsive to their requests, but increased the tax burden on the industry.[8] From 1925 as sugar prices fell the SNA stepped up its campaign to have taxation on the industry reduced and for more positive government support. A detailed report made by the *Comite de Defensa de la Industria Azucarera* in October 1928 recommended at least seventeen separate issues which were to be raised with the government.[9] These included tax reductions, improved credit facilities, cheaper guano, financial support for technical assistance programmes, and negotiations to get Peruvian sugar preferential treatment in the British market. Although it seemed at first that Leguía might be sympathetic to at least some of these requests, in the end virtually nothing was done. It has been suggested,[10] that the *hacendados's* lack of political muscle in this period was linked to the decreasing importance of sugar within the economy. While this obviously did them no good, it was the political defeat of the oligarchy by Leguía in 1919, when the economic fortunes of the sugar producers were at their height, that deprived them of a more decisive voice in government policy.

It is doubtful whether even if sympathetic government could have offset the effects of the continued slide of world sugar prices, but except for the willingness to repress all manifestations of labour unrest, Leguía's policies on balance tended to weaken the sugar producers' position. This was apparent both with respect to national economic policy and the control of the apparatus of local government, which being in the hands of Leguía's placemen, denied many *hacendados* their former untrammelled authority over key questions such as land or water rights disputes.

Leguía's downfall ushered in a period of extreme political turmoil.[11] Peace was not restored until General Oscar Benavides came to power in 1933. He was broadly sympathetic to the needs of the sugar producers, and it can be argued that in the 1930s the oligarchy regained something of the political power which it lost during Leguía's reign. Nonetheless, the social and political structure had been altered considerably and as Dennis Gilbert has observed,[12]

> The patrician democracy of the Civilista years (1895–1919) had been liquidated to be replaced by a more repressive type of government needed to deal with the higher level of political

mobilization after 1930. The oligarchy no longer ruled directly, but promoted and guided military governments which could be depended upon to protect its privileges.

How then did the state respond to the demands of the sugar producers? One of the most pressing problems, especially in the early 1930s, was widespread labour unrest. This had been sparked off by wage reductions and the attempt by the *hacendados* to increase the intensity of work,[13] and was organized by the newly formed Communist and *Aprista* parties. When Sanchez Cerro returned to the presidency in October 1931, his attitude towards the workers was unequivocally repressive, both before, and with greater ferocity after, the dramatic Aprista uprising in Trujillo in June of 1932. Overall, the sugar producers were less pleased with Benavides' handling of labour problems. Initially he adopted a policy of 'peace and concord' with APRA, but as this resulted in renewed political and union agitations by the Apristas, as well as opposition from the right, he soon returned to a more conservative stance.[14]

However, Benavides' strategy of combating APRA and maintaining relative internal stability did not rely entirely on force. He also instituted a series of modest social welfare measures.[15] When at one stage sugar growers complained about a government proposal to supply sugar to the domestic market at a fixed price, Benavides reminded them that '... these measures of *'benefico popular'* had been taken by his government in order to combat Aprismo, which takes advantage of these crises to act against public order'.[16] Whatever the methods used, both APRA and the unions were effectively suppressed under Benavides. This was of inestimable value to the sugar producers, for as will be shown, lower wages and greater control over labour were the chief weapons in combating the crisis of the 1930s.

The administrations after 1930 were also more responsive than Leguía had been to specific requests from the industry. Soon after his overthrow a law was passed suspending seventeen local export taxes on agricultural goods, and in 1934 the 'Assistance to the Sugar Industry Law' was passed. However, the law proved to be of little real assistance. Of greater importance was the formation of the *Banco Agricola* in 1931 to provide short–term loans to agriculturalists.[17]

Although Benavides was generally supported by and in turn supported the sugar producers, all was not to the latter's liking. They fought against the government's modest social reforms, continued to complain about various matters, and were unhappy when the President failed to use his discretionary powers under the 1934 Law.[18] The only partial success of the sugar interests in influencing government policy may have been due to divisions within their own ranks. For example, it was claimed that the two largest growers, W.R. Grace and the Gildemeister family[19] were being too cautious in their dealings with the government and '... these sugar producers give very weak support to the efforts made by the national producers, shielded because they are foreigners. In short, there is no unified action which could be a major force.'[20] It is unclear to what extent this situation actually undermined the political influence of the industry, but it is abundantly clear that when compared to most other major sugar producing countries, the Peruvian government did relatively little for its growers.

Intra-industry competition

While the crisis led to a degree of cooperative efforts among growers with respect to obtaining help from the state, within the industry a fierce battle for markets developed for the first time, and this was to change radically the structure and distribution of coastal sugar production. There was some disputes over the nearby export markets of Bolivia and Chile, but the main area of conflict was the domestic market.

Until prices began to slide in the mid–1920s most sugar producers had shown little interest in the Peruvian market. This now began to change, especially with respect to Lima. This had been supplied primarily by the relatively small estates in the nearby Rimac and Carabayllo valleys, who despite having higher costs (because of inferior growing conditions and considerably higher wages)[21] than the northern producers had survived by specializing in the production of the higher quality 'blanca' (plantation white) for the metropolitan market.[22] This earned them a premium of about 12 per cent over the price of export quality sugar. Some northern growers, particularly Cartavio, were attracted by this and began to switch part of their output to blanca and invade the central valley market. By 1928 the estates here were in serious trouble, and by the mid–1930s all the *ingenios* in the Lima valleys had been driven out of production.[23]

The attack on the Lima growers was only part of the intense and widespread competition which raged in the home market during the thirties. In the Southern Sierra the small Tambo Valley ingenio of Chucarapi faced a strong challenge from the northern estates of Pomalca and Cartavio.[24] In the valley of Santa, the British owned estate of San Jacinto complained that other growers were underselling them in what they considered 'their' local markets, and further to the north Cayaltí voiced the same complaint.[25] It was only when export prices began to rise once again that the scramble for control of the domestic market ended. By late 1939, the government had to intervene, as they had done during and immediately after WWI, to force the industry to maintain domestic supplies at reasonable prices.[26]

During the interwar period Peru's sugar producers faced extremely difficult conditions in the international market. Access to the United States was effectively blocked until the very end of the period by the country's commercial policy and there were serious threats to her principal markets in Chile and Britain. With the world price of sugar falling inexorably and with little substantive help from the state, the sugar growers turned on one another and in the process many estates, indeed an entire region, were driven out of cane production. Whereas there had been 33 *ingenios* in 1922, by 1937 there were but 14. At the same time production had risen slightly (315,310 metric tons to 388,722 metric tons) and the area under cane had also increased(50,812 hectares to 53,472 hectares).[27] Therefore, despite the hard times and the many *ingenios* which had closed, the industry had continued to grow, albeit very slowly. This pattern of a few very large, mainly integrated milling–cane growing estates was to remain until the agrarian reform began in 1969.

But how had any estates survived these years given the appalling state of the market and the fact that prices were often lower than production costs? To understand this it is necessary to look more closely at the response of the individual estates and changes made in production methods, costs and productivity.

Changes on the estates

The estates which continued to produce sugar were faced with the urgent need to reduce costs, but technical innovation was costly and while changes were introduced, as will be shown, the extent of these changes was severely limited by financial constraints. It is possible to offer a rough indication of investment in new factory and milling machinery, but because the figures in Graph 6.1 include neither all foreign suppliers, all classes of equipment, (i.e. tractors or other field equipment) or locally produced goods, they understate investment. Nevertheless, the trend shown is supported by the qualitative information derived from estate documents.[28] It shows that in the immediate post–war years many of the estates were re–equipped. This phase lasted more or less until 1925 when the coast was bit by devastating floods and world sugar prices began to fall. With the more precipitous decline from 1930 imports of new machinery declined to such a low level as to suggest only spare parts were being purchased.

Although it is apparent that a good deal of investment in factories and mills was made in the early 1920s, there are no long–run series which could be used to assess changes in milling or fabrication in the interwar years. Estate records show that extraction ratios did improve markedly before 1930,[29] and national statistics show a slight overall improvement in the crude fabrication ratio (sugar per ton of cane), with a more marked rise in the valleys of Trujillo.(see Table 6.1).

Graph 6.1. Peruvian Sugar Machinery Imports 1920–1940.

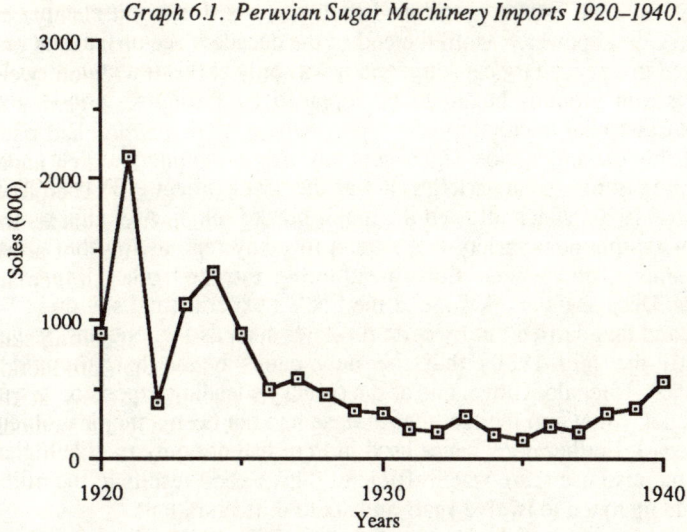

Source: Albert, *Essay*, 141a

There were continued modifications in processing, but it would seem that from 1925 the most important changes were made in the fields. There was the greatest scope for improvements here, for not only were operations more labour intensive than in the mills and factories, but also because agricultural costs accounted for by far the largest share of total production costs (about 60 per cent).[30] The desire to improve and mechanize field processes went well beyond wanting simply to reduce direct production costs. During and immediately after WWI and again in the early 1930s there was massive labour unrest on the coastal estates,[31] and this convinced many *hacendados* that any measures to reduce the number or workers would give them greater control and thereby reduce the possibility of trouble. Moreover, the large concentrations of workers and their families living on the estates meant considerable overhead expenses for the estates, such as housing, sanitation, recreation, health care, etc., and they sought to reduce these costs whenever possible.[32]

The first field process to be improved was cultivation, essentially weeding or the cleaning and remaking of furrows. From about WWI lighter, more efficient cultivating implements were introduced. This allowed *hacendados* to replace oxen with mules, the latter which were cheaper to feed, worked faster and did less damage to the furrows. From the early 1920s some estates began to replace animals with light tractors such as the Fordson. Although it was claimed that a tractor could do the work of 17 teams, because of problems with reliability and high repair costs in the early 1930s some estates abandoned their tractors, finding it cheaper to return to animal power.[33] Until the end of the decade it seems that tractors were used in a secondary capacity, and it was only at this time that the old Fowler steam ploughs began to be replaced by the more mobile and efficient Caterpillar tractors.[34]

Probably the single most significant advance in the interwar years was the planting of new cane varieties. It was the development of POJ 2878 in Java in the 1920s which allowed the major breakthrough. Although by the early 1930s this new variety had spread to many regions, on the whole Peru seems to have been slow in adopting this and other improved varieties. Despite the work done at the SNA's experimental station at La Molina and that carried out by certain estates such as San Nicolas, it was not until the mid–1930s that the new canes began to gain wider acceptance.[35] Gerado Klinge, one of the country's leading experts on sugar, put this delay down to the fact that disease had not been a major problem on the coast. Furthermore, canes used in Peru had not only to give higher yields, but also use scare water efficiently, give good results in the mills, and stand up to ten to twelve years and six to thirteen rattons.[36]

What precisely was the impact of technical change on the productivity of the sugar industry in these years? This is a difficult question to answer for the official statistics provide the only indication, and because of the way in which they were compiled (estate returns to the SNA) and that figures were aggregated by regions they do not offer an entirely reliable or sufficiently detailed picture. Furthermore, the collapse of most of the ingenios in the central valleys in the early 1930s means that subsequent national results may have improved simply because of the absence of these generally less efficient estates. Bearing all these problems in mind, the figures in Table 6.1 suggest that there was quite spectacular improvement in the interwar years. But it was a specific improvement. Land productivity (column 1) did not improve. In fact, it declined by about 0.2 per cent a year before the higher yielding varieties came into general use by 1940. This may have been due partly to the fact that the massive expansion in cane acreage in the provinces of Trujillo and Chiclayo (between 1918–20 and 1935–37 by 54 per cent and 75 per cent respectively) was only possible by moving onto less productive land.

The factory results (column 2) do show an important gain of 15 per cent, meaning that for very ton of cane an extra 16 kilos of sugar was being obtained by the 1930s. The most spectacular improvement was in labour productivity, in both the field and the mill.[37] This undoubtedly saved the industry from complete collapse. The simple equation given below allows a clear decomposition of the rate of growth of labour productivity.[38] It is, however, impossible to say how much improved

Table 6.1. Peruvian Sugar Industry: Measures of Productivity.

	(1)	(2)	(3)	(4)	(5)	(6)	(7)	(8)
National Data								
1918-20	107.5	10.7	11.5	10.9	65.0	.945	1.13	122.0
1928	109.0	11.8	12.8	11.8	80.6	.970	1.14	117.8
1935-37	103.2	12.3	12.7	16.2	102.6	1.307	1.55	156.4
1940-42	124.0	11.5	14.2	-	-	-	-	-
Chiclayo (Valleys of Lambayeque and Zaña)								
1918-20	108.5	9.8	10.6	12.9	72.8	1.219	1.48	160.9*
1935-37	94.1	11.1	10.5	13.4	95.6	1.284	1.49	139.7
1940	118.1	10.7	12.7	-	-	-	-	-
Trujillo (Valleys of Chicama and Santa Catalina)								
1918-20	110.0	12.5	13.7	12.6	93.2	.886	1.04	117.0
1935-37	105.8	13.7	14.5	19.8	133.4	1.419	1.64	170.1
1940	133.4	12.8	17.1	-	-	-	-	-

Note: (1) Cane (tonnes)/hectare cut; (2) Sugar/Cane per cent; (3) Sugar (tonnes)/hectares cut; (4) Sugar (tonnes)/worker; (5) Sugar (tonnes)/ mill worker; (6) Hectares cut/worker; (7) Hectares cut/field worker; (8) Cane (tonnes)/field worker. - means data not available. * This figure is obviously incorrect and reflects an underestimation of the number of workers. This means that the increase in labour productivity is understated.

Source: Peru, Ministerio de Fomento, *Estadística de la Industria Azucarea*, Lima, various years.

labour productivity owed to mechanization, was the result of better organization of working practices, or the greater intensity of work. The data indicate that there were substantive gains made, particularly in the province of Trujillo, and that improvement was not caused simply by the loss of less productive estates further south. Finally, the cost data available for individual estates generally support the trend indicated by the national statistics.[39] From the mid–1920s to the late 1930s Cayaltí's real costs fell by more than 30 per cent and on the extremely large Casa Grande estate real production costs were cut by almost 60 per cent. Given that the level of investment throughout the 1930s seems to have been so low, how were costs reduced and productivity increased?

(1) $\frac{d \log (Y/L)}{dt} = \frac{d \log (Y/L)}{dt} + d \log (A/L)$

(2) sugar/worker = sugar/ hectare cut + hectares cut/worker
2.57 per cent/year = 0.45 per cent/year + 1.817 per cent/year

(3) cane milled/field worker = cane milled/hectare cut + hectares cut/field worker
1.403 per cent/year = –0.227 per cent/year + 1.790 per cent/year

As has been argued, even without substantial imports improvements continued to be made during the 1930s. Gustavo Aspillaga claims that the severe crisis led to greater technical cooperation among the *hacendados*, as well as forcing each estate to look more closely at every sub–process in order to streamline working practices.[40] The changes which resulted could have had a cumulatively powerful impact on productivity. This is not an uncommon phenomenon, for as Salter has observed, '... many experienced observers rate the cumulative effect of small unnoticed modifications and improvements as equally great as the more significant changes generally regarded as innovations.'[41]' There is also evidence that higher productivity and lower costs can be attributed to a more intensified work regime and lower wages. For example, the official statistics show that wages fell generally in the 1930s in all regions except the valleys of La Libertad and Santa. But the estate records show that the official returns from these areas were falsified, possibly because of the widespread labour unrest at the time, and wages were clearly being cut. For example, in 1930 Casa Grande, Cayaltí, and San Jacinto all lowered wages by between 10 and 20 percent.[42] This move was undoubtedly followed by other growers, as *hacendados* in different valleys exchanged information on wages, and in Lambayeque they tried to agree a common policy on wage reductions.[43] In 1932 yet another round of wage cuts was initiated,[44] and this was followed two years later by further reductions.[45] During these years there were also attempts to intensify the working day. Although more research is needed on this point, the following letter is strongly suggestive of such moves. In July 1930 the Aspillagas wrote to their field manager,[46]

> We repeat to you the special recommendation of making an effort to increase the quantity of work that today is represented by the standard *tarea* in diverse works at Cayaltí. Effectively that a change in price or value of the daily wage is at the moment very difficult to make the sugar estates have to lengthen the *tarea* to a reasonable limit. One of the managers of Casa Grande has told us that on their haciendas they have arrived at a 15 per cent increase in the *tarea*.

Conclusion

That it survived at all in the face of such severe price falls shows that generally the Peruvian sugar industry responded reasonably well to the crisis of the interwar period. However, in the process the relative importance of sugar in the economy and the industry's profile were radically altered. In the 'dance of the millions' years after WWI (1919–21) sugar accounted for 34 per cent of Peru's export earnings.[47] By 1929 this had fallen to 12 per cent and was down to only 9 per cent by the years 1937–39. Although output rose over the period, performance was sluggish, production rising by only 1.5 per cent per year, while the annual increase in the volume of exports was an almost imperceptible 0.29 per cent. At the same time the value of sugar exports declined by almost 5 per cent per year, and by the late 1930s was less than 40 per cent of the post-war level. The crisis also saw a dramatic restructuring of the industry, with the number of *ingenios* being more than halved. Sugar production in the central valleys, which accounted for 30 per cent of output before WWI, was completely decimated, and a major prewar trend continued as the industry became concentrated in the northern departments of Lambayeque and La Libertad.

The estates which weathered the years of crisis tended to be large, those in the better growing areas of the North, haciendas like Cartavio or the San Jacinto which were foreign owned, or those controlled by the Gildemeisters, a Peruvian family which had strong foreign links, both with Germany and Chile. But while size, area, foreign connections, or access to special markets (as in the case of Chucarapi) may be important in explaining why certain estates survived as sugar producers, it was the phenomenal increase in labour productivity and the lower costs which derived from both this and wage cuts which allowed the industry as a whole to survive. Innovation was, of course, a major part of the process, but it is clear that, as in all capitalist crises, it was the workers who had to shoulder the most onerous burden. They did not do so passively. The reduction of wages and the increased pressure on the sugar workers to produce more contributed to violent political and social unrest brought about in Peru following the Crash of 1929. In the early 1930s Haya de la Torre's Aprista Party began to organize sugar workers on many estates, a bitterly contested national election was followed by a bloody and abortive revolt by the Apristas in Trujillo in July 1932 and then the effective suppression of both APRA and the workers' organizations.[48] The willingness of the state to help the *hacendados* control labour was undoubtedly a key factor in the industry's ability to survive these difficult years. In the end it outweighed the government's failure to subsidize sugar growers or conclude preferential treaties.

References

1. V.P. Timenoshenko and B.C. Swerling, *The World's Sugar. Progress and Policy*, Stanford, 1957, 158.

2. *La Vida Agricola* (Lima), Feb. 1926, 153; Sept. 1927, 753–54.

3. C.P. Kindleberger, *The World in Depression 1929–1939*, London, 1973, 190–91.

4. For details of treaty see, Bill Albert, 'Sugar and Anglo–Peruvian Trade Negotiations in the 1930's', *Journal of Latin American Studies*, 14, 1, (1982), 128–32.

5. Timoshenko and Swerling, *World's Sugar*, 201–207.

6. See Albert, 'Anglo–Peruvian', 132–42.

7. *Ibid.*, 140.

8. *SNA Mss.*, 'Azúcar Legislación Tributaria', Lima, Sept., 1937. (Archivo del Fuero Agaria, Lima). *La Vida Agricola*, Dec. 1930.

9. *SNA Mss.*, 'Informe del Comite...', Oct. 1928.

10. Rosemary Thorp and Geoff Bertram, *Peru 1890–1977. Growth and Policy in an Open Economy*, London, 1978, 49.

11. Frederick Pike, *The Modern History of Peru*, London, 1967, 250–68.

12. Dennis Gilbert, 'The Oligarchy and the Old Regime in Peru', unpublished PhD thesis, Cornell University, 1977, 97.

13. See below,

14. Gilbert, 'The Oligarchy', 116–17.

15. Julio Cotler, *Clases, estado y nación en el Perú*, Lima, 1978, 252.

16. *Cayaltí Mss.*, Cartas Officiales, Lima–Cayaltí May 20, 1937 (Archivo del Fuero Agraria).

17. SNA, *Memoria de ... 1932–1933*, 14–15, 193–94, 216–21.

18. SNA, *Memorias de ... 1935–36*, 11–13, *Memorias de... 1936–37*, 121–132, *Memorias de ... 1938–39*, 94–98.

19. By 1930 these two groups were producing about 50 per cent of the country's sugar. Grace was a large multinational firm, while Gildemeister, although considered by many as foreign, was in fact a Peruvian based company, established by German immigrants.

20. *Cayaltí Mss.*, Cartas Particulares, Lima–Cayaltí, Nov. 8, 1935.

21. Bill Albert, *An Essay on the Peruvian Sugar Industry 1880–1920 and the Letters of Ronald Gordon, Administrator of the British Sugar Company in Cañete*, Norwich, 1976, 140a–155a.

22. *SNA Mss.*, Letter from G. Allen R to President Comite Defensa del Azúcar, Oct. 22, 1928.

23. *La Vida Agrícola*, April 10, 1936.

24. *Chucarapi Mss.*, Letters between Chucarapi and E. de la Pierda, Aug.–Sept. 1938 (Archivo del Fuero Agraria).

25. *San Jacinto Mss.*, Corres., San Jacinto–Lima, Dec.5, 1932 (Archivo del Fuero Agraria); *Cayaltí Mss.*, Lima–Cayaltí, May 17, 1936.

26. SNA, *Memoria de... 1939–40*, 14.

27. Peru, Ministerio de Fomento, *Estadística de la Industria Azúcarera... 1922.*; *Estadística...1937.*

28. For a more detailed treatment of technical change see, Bill Albert, 'Causes del cambio technológico en la industria Azúcarera peruana, 1860–1940', in *El Azúcar en America Latina y el Caribe*, ed. Horacio Crespo (forthcoming).

29. *Ibid.*

30. *Cayaltí Mss*, Cartas Particulares 1934–36, March 18, 1935.

31. Albert, *Essay*, 178a–218a; Michael Gonzales, *Plantation Agriculture and Social Control in Northern Peru, 1875–1933*, Austin, 1985, 171–188.

32. *Cayaltí Mss.*, Cartas Reservadas, Lima–Cayaltí, Dec.6, 1926.

33. *San Jacinto Mss.*, Letter Jan. 20, 1932.

34. *Cayaltí Mss.*, Cartas Particulares, Lima–Cayaltí, June 25, 1935; Cartas Officiales, Lima–Cayaltí, Sept.22, 1937; R.J. Lockett, 'Records of the Nepeña Valley in Peru', *Revista de Valle Nepeña,* 1958/59, no.4.

35. *San Nicolas Mss.*, 'Memoria de la Administración 1932–1933', Cuadro no.4; Report, Jan. 20, 1932.

36. *La Vida Agricola*, May 1936, 416; Jan.1943, 69–73.

37. The improvement was probably greater than suggested by the figures because of the underestimation of field workers in the province of Chiclayo in the first period. See note to Table 6.1. This is a crude indicator as what is needed is hours worked, not simply the number employed.

38. I have to thank Chris Scott for this suggestion. The figures in the equations are not exact because of the problems of averaging, but they do give a reasonable indication of the relative proportions of land and labour productivity.

39. All cost data from Albert, 'Cambios...'.

40. Interview, Lima, March 1974.

41. W.E.G. Salter, *Productivity and Technical Change*, Cambridge, 1969, 5.

42. *San Jacinto Mss.*, Actas del Directorio no.2, Sept.22, 1930; *Cayaltí Mss*, Cartas Particulares, Lima–Cayaltí, Aug. 8, 1930.

43. *Cayaltí Mss*, Cartas Particulares, Lima–Cayaltí, Aug.8, 1930.

44. *Cayaltí Mss.*, Cartas Particulares, Lima–Cayaltí, May 27, 1932; *Casa Grande Mss.*, Annual Report, June 30, 1932; *San Jacinto Mss.*, Actas del Directorio, June 10, 1932; Alica Suarez Guimarey, 'Bases Materiales para el desarrollo del sindicalismo en Lambayeque 1930–1960', unpublished Mss., Lima, 1977/78,18–20.

45. A. Suarez G., 'Bases...', 20–1.

46. *Cayaltí Mss.*, Cartas Particulares, Lima–Cayaltí, June 12, 1930.

47. Albert, *Essay*, 30a.

48. Peter Klarén, *Modernization, Dislocation and Aprismo. Origins of the Peruvian Aprista Party 1870–1932*, Austin, 1973, chapter 7.

7

The Cartelization of the Mexican Sugar Industry, 1924–1940

Horacio Crespo

Although political problems were to continue, the election of General Alvaro Obregón as president in 1920 ended the most intense period of armed struggle of the Mexican Revolution. From then until 1928 the main problem facing the country was the difficult task of establishing new political, social and economic norms to accommodate the various sectors that had participated in the Revolution. In the next period to 1934, the still weak Mexican economy, under the strong conservative hand of President Plutarco Elías Calles (1924–28) and his chosen successors was badly hit by the world depression. Partially in response to this crisis, in the years 1934–1940, there was a radical shift in government policy. Lazaro Cárdenas pushed through a national project with a strong emphasis on both substantive agrarian transformation and laying the basis for industrial growth and state ownership of the country's basic resources. This project was based on popular mobilization through corporative organizations subordinate to the state.

This is the general political background against which the important changes in the Mexican sugar industry in the interwar period occurred. The profound crisis of the 1930s brought a managerial response of cartelization which could only work with increased state involvement. This was a move entirely in sympathy with Cárdenas's national project and was to culminate four decades later with the almost complete nationalization of Mexican sugar production.

Sugar production, 1920–1940

In the two decades before the Revolution the country's sugar production had increased three–fold. In these years the processing sector had been substantially modernized and cane acreage increased. This process was ended by the outbreak of armed conflict. Production which had reached about 155,000 metric tons in 1911/12, was down to only 44,000 by 1917/18.[1] As shown in Table 7.1, when peace was restored production quickly reached previous levels. However, regional distribution was now considerably altered, there had been a greater concentration into larger units of production and the composition of the most important group of sugar producers had also changed.

In the state of Morelos, which during the pre–revolutionary years was the country's leading producer (about a third of total output), sugar production had been virtually eliminated by 1921. This was largely because the state was the centre of the peasant–based Zapatista Movement. The bloody conflict, which had ravaged Morelos for a decade, left all 26 *ingenios* either burned down or with machinery destroyed or stolen.[2] Furthermore, cane lands had been abandoned and valuable irrigation works left to deteriorate. But probably the most important factor affecting the state's sugar industry was the political compromise worked out between central government and the Zapatistas, which resulted in an energetic agrarian reform being set in motion here. This did not happen in other cane–producing states, where the process was delayed until the Cárdenas Administration. In Morelos the *haciendas* lost a great deal of their irrigated lands to peasant co–operatives (*ejidos*).[3] This meant that the old *hacienda* system, which combined processing and cane growing under the control of a single powerful *patrón*, could never be reestablished here. Although two or three small *ingenios* reopened in the late 1920s and early 1930s, the rebirth of the industry in Morelos did not begin until the formation in 1938 of the Central 'Emilano Zapata', a worker–peasant co–operative, in Zacatepec. This was to become of major national importance from the 1940s.

Another region badly affected by the upheaval of the Revolution was the southern valleys of the state of Puebla, on the border with Morelos. Here too Zapatistas had been active. Of the eight *ingenios* working in 1913, only four were not destroyed. The difference in Puebla was that the sugar industry staged a fairly rapid recovery because Atencingo, an old *ingenio*, was quickly transformed into a central mill.[4]

The major change in the regional structure of sugar production was the greatly increased importance of the state of Veracruz, mainly because of the elimination of its old competitor, Morelos. With its 22 *ingenios* in operation in 1922 it was the state with the largest number of producing units. These varied in size from 120 tons per year per mill to one producing 8,000 tons. The key to the development of sugar production in Veracruz was the vigorous expansion and modernization of two major *ingenios*, El Potrero and San Cristóbal. Accounting for 26.7 per cent of the state's output in 1922, they had increased their share to 45.7 per cent by 1929, and 51.9 per cent two years later. This level was more or less maintained during the 1930s but declined in the following decade as other *ingenios* modernized their plants.

Table 7.1. *Mexican Sugar: Production, Consumption, Imports, Exports and Reserves, 1920–1940* (metric tons).

Year	Output	Consumption	Imports	Exports	+/–	Reserves
1920	72500	–	6374	12310	–	–
1921	95800	–	18815	49	–	–
1922	155790	–	3454	6	–	–
1923	134700	–	492	9432	–	–
1924	159930	–	277	16240	–	–
1925	163420	154700	406	7346	1780	53096
1926	191940	173800	435	6634	11941	65037
1927	184050	190200	557	3633	–9226	55811
1928	167240	191000	421	16	–23355	32456
1929	180980	195800	2043	21	–12798	19658
1930	215600	205000	874	256	11218	30876
1931	262615	188300	189	26825	17679	78555
1932	228888	166300	62	6807	55843	134398
1933	187451	196580	62	86775	–95842	38556
1934	188245	221650	10216	51	–23240	15316
1935	266214	239530	476	192	26968	42284
1936	307646	269190	466	31	38891	81175
1937	278825	277070	17	79	1693	82868
1938	303376	303632	16	584	–824	82044
1939	331265	334883	21	5525	–9122	72922
1940	292195	358470	20	28	–66283	6639

Sources: Unión Nacional de Productores de Azúcar, *El desarrollo de la industria azucarera en México durante la primera mitad del siglo XX*, Mexico, 1950, 43. Banco de México, *La industria azucarera de México*, Vol.I, Mexico, 1952, 256–58.

The State of Sinaloa, in the northeast, became a prominent sugar producer after 1920 on the basis of two large *ingenios*, Los Mochis, the country's largest single producer during the interwar years, and El Dorado. From 7.8 per cent of Mexican sugar output in 1911/12, the state increased its share to 15.8 per cent in 1922, and 23.7 per cent in the period 1926–1930. Its importance continued to rise in the following quinquennium (29.7 per cent), but fell dramatically thereafter (to 20.1 per cent 1936–40), as the harvest of the two major *ingenios* remained static throughout the 1930s.

The other two important sugar–producing states were Jalisco and Tamaulipas. The former accounted for about 10 per cent of national production between 1926 and 1940. In Tamaulipas, it was the principal *ingenio* of El Mante which allowed output here to grow by the early 1930s to about the same, and in some years higher, levels than in Jalisco.

Despite the change in regional distribution of sugar production, the concentration of the industry in relatively few states continued. For example, in 1913 the five most important states, Veracruz, Morelos, Puebla, Sinaloa, and Michoacán, all producing over 10,000 metric tons, were responsible for 83.4 per cent of the nation's sugar. In the following five year period, with Tamaulipas replacing Michoacán, the figure reached 90.6 per cent. By the late 1930s it always stood at about 80 per cent or more. This degree of concentration was important in that it offered the possibility of forming a national producers' organization with the capacity to make decisions and to regulate the industry.

Another important structural feature of the Mexican industry, which was also clearly an international phenomenon as well, was the continued concentration of production in a few *ingenios*. In 1922, the country's six largest *ingenios* produced 23.6 per cent of total sugar. Two years later participation had risen to 38.2 per cent, jumped to 43.6 per cent by 1929 and reached a peak of 55.7 per cent in 1934. The fact that such a large proportion of production was in so few hands further increased the likelihood of successful centralized control.

The third major change in the post–revolutionary sugar industry was in ownership. The destruction of the industry in Morelos marked the complete elimination of the most powerful group of producers, whose greatest strength, and ultimately their greatest weakness, had been their close political and family links with the Diaz regime. In the 1920s a new group of sugar entrepreneurs began to form, who, over time, were to become immensely influential within the Mexican economy, not only because of their investments in sugar but also in that their expansion affected other key sectors, notably banking and finance. In political terms they also became powerful. Functioning as a pressure group, and often working as members of political and economic alliances, they had a major impact on government policies. They even seem to have had an influence on the final decisions taken by the government to nationalize the greater part of the *ingenios* in the 1970s, a move which marked the end of this group's power within the country.

Sugar crises and the cartelization of the industry

Throughout this century the Mexican sugar industry has faced various crises, the result fundamentally of the maladjustment of production with domestic demand. This was because the industry developed basically to supply the internal market, and from the 1870s technological modernization and an increase in the scale of production led to a tremendous growth in output which outstripped the much slower increase in consumption. Within this system exports served as a safety valve to maintain domestic prices.[5] The main problem with this was that Mexican sugar was grossly uncompetitive in the world market and producers, whose domestic market was protected by tariffs, faced large losses when forced to sell outside the country.

One way to combat crisis was to organize, and in the 1870s sugar producers in Morelos and Puebla had done so in order to raise domestic prices and encourage exports. Subsequently, there were other regional associations formed, and in 1903 the first national organization, the *Unión Azucarera,* was established.[6] After the most violent period of the Revolution had ended sugar producers once again sought to organize, and in 1919 many of the most important *ingenios* formed the *Cámara de Productores de Azúcar.* The purpose was to protect and promote the sugar industry. Among their specific proposals was to insure tariff protection, improve export opportunities, and obtain larger and better credit facilities. On the other hand they fought hard against agrarian reform, which was seen as a major danger to the integrity of the industry. However, the *Cámera* never became a true national representative of the entire industry, and in 1924 it was transformed into a cooperative society with limited responsibilities.[7]

In 1921, in Nogales, the Sonora Commission Co. was established.[8] It was a regional body, mainly for northern producers, which was set up to regulate surpluses, and apportion losses occasioned by the dumping of sugar abroad. It changed its name one two occasions and lasted until 1931. In 1925 another group, *Compañía Comercial Comisionista, S.A.,* was organized in Mexico City to sell sugar produced in Puebla and Veracruz and to divide the national market with the Sonora Commission during the grave crisis which hit the industry in the years 1925–27. The *Cámara de Productores* tried unsuccessfully to mediate between the two regional groupings. In 1926 the *Compañía Comercial* was wound up. At that time there was yet another body, *Agencia de Ventas del Sur*, also selling sugar from Puebla and Veracruz. Evidently, the situation of the central producers was much more anarchic than in the north. An indication of this was the defection of the *ingenio* El Potrero in Veracruz, which allied with the northern producers in the late 1920s.

The main problem facing the aforementioned organizations was that while the successive sugar crises were national in origin they were only able to respond on a regional basis. Instead of cooperation this tended to encourage regional competition, and this in turn adversely affected the entire industry. Another difficulty, both within and between regions, was that the producers were reluctant to absorb the losses on exports, especially when the world price was well below production costs. Furthermore, all these groupings were at best only able to sustain activity during times of overproduction. An extremely significant conclusion from all this for the Mexican sugar industry is that '... official pressure has been necessary, exercised through tax subsidies, to achieve a strong organization bringing together the entire industry and regulating production, sales and prices, under the direction and intervention of the government.'[9]

Table 7.2. Annual Average Sugar Mexican and World Sugar Price 1920–1940 (pesos per kilo).

Year	Mexico	World	Year	Mexico	World
1920	73.5	–	1931	19.5	13.3
1921	37	–	1932	17.5	9.4
1922	29	1.7	1933	25	10.3
1923	34	6.1	1934	26.5	11.1
1924	30.5	5	1935	25.5	10.7
1925	25	7.6	1936	25	10.9
1926	26	8	1937	28.5	14.1
1927	23.5	1.1	1938	28.5	12.2
1928	30	6.6	1939	28.5	17.0
1929	30.5	0.4	1940	28.5	14.2
1930	27.5	4.6			

Sources: Unión Nacional de Productores de Azúcar, *El desarrollo de la industria azucarera en México* , 51. Banco de México, *La industria azucarera de México*, Vol.I, 384, Note: World price London cif converted to New York cif.

The situation facing the industry in the mid–20s was extremely unfavourable. The major problem was that many *ingenios* were in serious financial difficulties, especially those in Veracruz. To make matters worse, rising output and falling exports led to a drastic price fall in 1927 (Tables 7.1 and 7.2). The extent of the crisis resulted in the first important intervention of the Mexican government, excepting import duties, in the affairs of the sugar sector. By a decree of August 30, 1927[10] an excise tax of 2 cents was imposed on each kilogram of sugar sold in the country. The money collected was to serve as a fund to subsidize exports. This would be paid to those producers who exported and joined together in a Producers' Society recognized by the Secretary of Industry, Commerce and Labour. At the same time a National Sugar Commission was created. This was an official body to oversee the industry's affairs. In December of the same year another decree established with greater clarity official sugar policy[11], which aimed to improve production through applying new techniques to agricultural methods, seed selection, the construction of central mills and the search for new industrial applications for cane. It was also decided that credit would be granted to *ingenios* through the *Banco Nacional de Crédito Agrícola*, an official institution, that the association of producers would be encouraged, and that the taxes on cane, sugar and alcohol for industrial use would be ended. Furthermore, it was agreed to urge the setting up of joint commissions of workers and sugar producers so that better understanding could be reached between the two sides. This last element was the work of Luis Napoleón Morones, the Secretary of Industry, Commerce and Labour and the leader of the most important workers' union in the country, the CROM.

This first official action, the intention of which was to force all the producers to get together in a single organization capable of confronting the problems besetting the industry, had an extremely short life. In April 1928, the tax was suspended and the association cancelled, as stocks fell and domestic prices rose (Tables 7.1, 7.2). Nevertheless, the experience was of great importance in that all the elements, except the worker–producer commissions, were to figure in the subsequent measures for organizing the industry. Of central importance was, of course, the regulatory and controlling roles assumed by the state.

But, during the 1920s there was another important aspect of government intervention which was to have a major impact on the sugar industry. In August 1927, President Calles decreed that lands in cane, together with those given over to coffee, cacao, alfalfa, vanilla or rubber, would thereafter not fall into the category of holding which would be subject to the agrarian reform. This move was entirely consistent with his agrarian politics, the fundamental objective of which was to create a solid class of agricultural proprietors and further to give security and stability to the sugar producers who felt threatened by the demands for agrarian reform made by the peasants. Some observers criticized this measure, maintaining that the President was defending his personal interests, in La Primavera and El Mante, and that he was siding with the *latifundistas*.[12]

After 1930 the sugar industry was once again in crisis, but this time it was extreme, probably worse than that experienced between 1904 and 1908. In 1930/31 production rose by 50,000 metric tons (22 per cent), and the accumulated surplus increased to almost 50 per cent of annual domestic consumption. To make matters worse, the latter fell appreciably between 1930 and 1932 because of the severe conditions affecting the entire economy. The export safety valve, which usually had helped resolve the industry's problems, exploited as never before between 1931 and 1933, was now extremely problematic, given that by the latter year the world price had fallen to less than half of the domestic price.

It was becoming apparent that the situation facing the industry was unprecedented. It seemed on the verge of collapse. In response to this some of the main sugar industrialists approached ex–President Calles to get the government to intervene in order to regulate production and guarantee domestic prices. Calles, known as *el jefe máximo*, was the obvious choice for this representation because of his substantial interests in the sugar industry and because he continued to dominate and direct the Mexican political system. He advised recourse to a loan from the Banco de México to finance the losses incurred on exports necessary to clear the surplus sugar. He also put forward the idea of yet another association of sugar producers, part of the legislative programme of his own government in 1927. Calles interceded on behalf of the sugar industry with the government of his virtual appointee–president, Ortiz Rubio, drawing on the assistance of the Secretary of Industry, Commerce and Labour, Aarón Sáenz, who also just happened to be the most powerful and influential sugar producer in the country.

On January 3rd, 1931 the federal government instituted its 1927 strategy, but in a more severe form.[13] A 5 *centavos* tax was imposed on each kilogram of sugar produced. The same amount, less administration costs, were returned to all those producers who joined the *Compañía Estabilizadora de Azúcar y Alcohol S.A.*, a commercial company set up to export the warehoused sugar surplus and impose controls to limit production, controls established by the *Comisión Estabilizadora de la Industria Azucarera*. This latter group comprised representatives from the government and the producers. Eight of the largest sugar producers together with the *Banco Nacional de Crédito Ejidal* were shareholders of the *Compañía*.[14]

This move was effective in the the medium term. Between 1931 and 1933, 112,000 metric tons of surplus sugar were exported, which substantially reduced reserves. At the same time production was also cut by 75,000 tons. All this in turn led to a increase in domestic prices. The *Banco de México* opened a 12 million *peso* line of credit over a four year period, which allowed the losses on exports to be recouped. Under the 1931 decree the producers agreed to deliver 20 per cent of their production to the *Comisión Estabilizadora* in order to reduce surpluses and it was also established that the payment of the tax could be made in cane not used for making sugar or alcohol, although it could be used as fodder.

Despite the eventual success, in mid–1931 the difficulties facing the industry seemed to worsen because of the 1930/31 harvest, the most abundant ever. In July a meeting of producers was held to discuss stabilization plans. The most pressing problem facing the new *Compañía Estabilizadora* was the continued competition between the producers in Veracruz and those in Sinaloa. Aarón Sáenz, sugar baron and the government minister directly responsible for this problem, observed that the government intervention would not go so far as to protect the entire sugar industry. He said, 'The Government believes that outside of this action (the decrees) its intervention in the administrative, economic and financial life of the industry would not be beneficial, in that it would create artificial situations which would hinder the free play of economic factors, leading to disequilibrium and creating worse problems than those it was attempting to solve.'[15] In this way the manner in which the crisis was handled was to be hardest on the small and medium sized producers, while opening up new possibilities for the larger ones. It is not surprising that just when the crisis was at its worst, the *Ingenio El Mante* , of which Sáenz was part owner, increased its production. At the same time many others were bankrupted or put in severe financial difficulties.

The result of the July meeting was that the *Compañía Estabilizadora* was wound up and a new association was established. This was *Azúcar S.A.*. It had a capital of 100,000 *pesos* and all 79 of the country's *ingenios* were made members, except a number of small ones in the states of Tabasco and Yucatán.[16] Its leading positions were assumed by the most powerful sugar producers. Aarón Sáenz himself, after resigning from the government in January 1932, was made general manager. Such leadership insured that at least the new company would be in a position to discuss disputes and arrive at binding decisions.

The new enterprise functioned as the marketing agent for the entire sugar industry. All the associated *ingenios* were obliged to sell to it all their sugar, alcohol, etc. which was then sold in common. *Azúcar S.A.* sold the various products at prices authorized by the government, being careful that surplus stocks were such as to insure national demand and maintain an stable price. In the case of excess production and the need to export, permission had to be obtained from the government. The company was also put in charge of imports, if and when this might prove necessary. Because the company was in effect a monopoly, a ruling was required on article 28 of the Constitution, which prohibited monopolies. It was argued that *Azúcar S.A.* was not a speculative enterprise, but one which sought to regulate prices in the interests of both consumers and producers.

The setting up of *Azúcar S.A.* finally resolved many of the most seemingly intractable problems which had long beset the Mexican sugar industry. It also allowed the industry to overcome probably the most severe crisis in its history. The cartel which was established finally ended the brutal competition between producers and replaced it with domestic market quotas and the acceptance of shared losses on exports. The great sugar barons had, therefore, looked to the state to solve a problem which they had let get completely out of hand.

But marketing was not the industry's only difficulty. Finance too was a constant worry. This was partially alleviated through the advances on sales given by *Azúcar S.A.* to the *ingenios*. But more important was the government subsidy which helped establish the *Banco Azucareros, S.A.* in 1932. This eventually became the *Banco de Industria y Comercio* and finally the *Banco Confía, S.A.*. Although initially concerned with the sugar industry, this bank and its successors were to extend their interests into other economic spheres. In this way the sugar producers were to become one of the most important banking powers in the country.

On August 22, 1938 the sugar cartel was once again reorganized, as the *Unión Nacional de Productores de Azúcar, S.A.* (UNPASA).[17] This was to bring it in line with the new legal dispositions dictated by the Cárdenas government. Essentially this meant that representatives of various ministries were placed on the administrative council and given voting rights here and in the association's general assembly. By these means the Mexican state undertook clear leadership and control of the country's sugar industry, making this position even more apparent when in 1937 it legally sanctioned a collective labour contract which was to affect all the nation's sugar workers.

Conclusions

The period 1920–1940 was one in which the modern structure of the Mexican sugar industry was firmly established. This encompassed the modernization of both technology and productive relations in an industry which in these years became fully capitalist. In what may seem as something of a contradictory trend, the later part of this period also saw increased government intervention, an intervention which was to result in the virtual state control of the most important decisions affecting levels of production, marketing, finance, pricing and credit.

During its period of armed struggle the Mexican Revolution profoundly affected the country's sugar industry. The most important result was the elimination of dominant *ingenio–hacienda* complexes in the principal producing centre of Morelos. After 1920, the leadership of the industry passed to producers in the states of Veracruz and Sinaloa. But it was not only during the era of violent conflict that the industry felt the impact of the Revolution. The changes experienced after 1920, were in many respects more profound and far reaching.

It can be argued that the political settlement of the Revolution was clearly reflected in the experience of the sugar industry. On the one hand in Morelos, where the Zapatistas remained strong, the implementation of agrarian reform literally cut the ground from under the old patronal estates and made it impossible for the sugar industry to re-emerge in its old form. The formation of farming collectives (*ejidos*) here might be seen as the more socially progressive face of the Revolution. At the same time, it was the emergent sugar capitalists, not the peasants or the *ingenio* workers, who came to dominate the new sugar industry. These capitalists can also be seen as 'progressive', in that their role was to eliminate the traditional, backward elements within the sugar industry and accelerate the capitalist transformation, already in progress before 1910.

An important aspect of this transformation was the gradual separation of the three principal interests within the industry, the industrialists, the workers and the cane farmers. This process was significantly hastened by the agrarian reform, which made the integrated agro–industrial estates, such as those in Morelos, impossible to maintain. But this change did not come immediately, in most parts of the country it had to wait until Cárdenas came to power in 1934.

Although the state did not become directly involved in the process of production, it became the decisive force within the industry. Without its intervention the cartelization and resultant stabilization of the industry would have been impossible. The price paid by the private sugar interests was to set the stage for their own subordination. Although for many years this was to also be economically extremely favourable for them, in the end it resulted in the loss of their economic and political power when the industry was fully nationalized in the 1970s.

References

1. Unless otherwise noted, all sugar statistics in this work taken from: Unión Nacional de Productores de Azúcar, *El desarrollo de la industria azucarera en México durante la primera mitad del siglo XX*, Mexico, 1950, 43.

2. Revista Industrial de México, *Directorio de la Industria Azucarera de México en el año de 1925*, Mexico, 1925, 68.

3. Carlos González Herrera y Arnulfo Embriz Osorio, 'La Reforma Agraria y la desaparición del latifundio en el Estado de Morelos', in H. Crespo, (ed.), *Morelos. Cinco Siglos de historia regional*, Mexico, 1984, 285– 98.

4. *Directorio de la industria*, 66, David Ronfeldt, *Atencingo. La política de la luncha agraria en un ejido mexicano*, Mexico, 1975, 19.

5. H. Crespo, 'La industria azucarera mexicana y el mercado externo, 1875–1910', in H. Crespo and S. Manigat (eds.), *Oro blanco y capitalismo*, (forthcoming).

6. H. Crespo, 'El azúcar en el mercado de la ciudad de México, 1885– 1910', in H. Crespo, (ed.), *Morelos, 188–191*.

7. *Directorio de la industria*, 19– 20.

8. On the successive sugar organizations since the Comission Co. and the foundation of *Azúcar S.A.* and U.N.P.A.S.A. see, Unión Nacional de Productores de Azúcar S.A., *Memoria de los primeros cincuenta años de UNPASA*, México, 1981, 9–55; 'Breve historia de la organización de la industria azucarera en México', in *Azucareros de México*, Sept. 1960, 21– 2; 'U.N.P.A.S.A.: Su Origen y desarrollo', series of articles in, *Boletín Azucarero*, II, 9–17, April–Nov., 1950; Banco de México, *La industria azucarera de México*, Vol.I, Mexico, 1952, 44–6, 51–3.

9. Banco de México, *La industria azucarera*, Vol.I, 44.

10. *Diario Oficial*, Aug. 31, 1927, XLIII, no.52.

11. *Diario Oficial*, Dec. 7, 1927, XLV, no.31.

12. Emilo López Zamora, *La situación del Districto de Reigo de El Mante*, Mexico, 1939, 18.

13. On the crisis of 1930–31 see, Aarón Sáenz, *Memoria de la Secretaría de Industria, Comercio y Trabajo*, Mexico, 1931, 347–82.

14. *Memoria de los primeros cincuenta años* , 18.

15. Sáenz, *Memoria* , 353.

16. *Memoria de los primeros cincuenta años* , 208–210.

17. *Ibid.*, 26, 111.

8

The Cuban Sugar Economy in the 1930s

Brian Pollitt

In the 'overproduction crisis' which first became apparent from 1925, Cuban nationalists foresaw the ruin of the small–scale agricultural and industrial sector of the sugar economy in which Cuban interests were concentrated. The strongest candidates for survival were judged to be the more modern, large–scale agro–industrial sugar complexes owned or controlled by US interests. The most pessimistic forecasts underestimated the extent of the crisis which developed in the early 1930s, but surprisingly, it was the technologically and financially weaker Cuban sugar interests which emerged from the period strengthened at the expense of their stronger US counterparts. In this paper we will consider the nationalist 'project' for the survival of the Cuban sugar economy from its evolution in 1925 to its implementation in the politically turbulent years of the Great Depression.

In a context of comprehensive US domination, the most distinctive feature of Cuba's economic history from 1900 was the extraordinary growth of sugar production. It multiplied fivefold, from one to five million tons, between 1904 and 1925. By the latter year Cuba accounted for 23 per cent of total world output.[1] The bulk of the increased production came from the eastern provinces of Camagüey and Oriente. From only 200,000 tons (or 20 per cent of national output) in 1904, these regions increased their production to 2.8 million tons (53.5 per cent) by 1925.[2] As will be shown, this aspect of the sugar industry's development profoundly influenced both the nature and resolution of central conflicts during the 1930s. The opening up of the sparsely cultivated and populated hinterlands of these provinces was associated with the construction of large–scale milling complexes which developed as apparently unassailable fortresses of foreign modernity. These were to become the object of a sustained and partially successful assault in the battle for the survival of the 'native' sugar industry in the Depression years.

97

The primary stimulus to this expansion of the Cuban sugar industry was the rapid growth of the US market to which Cuban sugar had preferential access. There was also massive US investment in the island's economy. Direct US investment was estimated at $200 million in 1906. By 1927 it had reached about $1 billion. Over 60 per cent of this was directly or indirectly deployed in the sugar economy in mills, railways, or land[3], although some of this must have been accounted for by US banks foreclosing on local mills after the Dance of the Millions boom collapsed in 1921. For example, it was reported that the National City Bank of New York alone took possession of 50 to 60 mills in the summer of that year, although not all remained in the bank's hands.[4] Jenks estimated that in 1926–27 the 84 US owned mills out of 177 grinding accounted for over 60 per cent of total output. Furthermore, they were judged to own or lease 22 per cent of the national territory, comprising more than 40 per cent of the total area in farms.[5]

While much criticism was subsequently to be made of the extent and nature of this US domination of the Cuban economy, the bulk of the land controlled by the large foreign companies had been previously uncultivated. Moreover, there is also no evidence of proletarianization of large numbers of independent farmers as a consequence of the growth of enterprises; the number of *colonos* seems to have increased rather than fallen between 1899 and the mid–1920s. In general, employment and real income for the majority of Cubans increased over the first quarter of the 20th century, and the burden of national criticism was aimed not so much at the US presence, as at Cuban decadence and cultural decline.[6]

Although US capital dominated Cuba, the sugar interests it represented were outgunned in the US Congress by mainland producers and those from the insular territories of Hawaii, Puerto Rico, and the Philippines. This became evident in 1920 and 1921 with the changes in tariff policy and was to become painfully clear in 1930 with the imposition of the Smoot–Hawley Tariff. To its lack of power in the US market was added its very limited success in encouraging other major producers to restrict output.

The early thirties were disastrous years for the Cuban sugar industry. In 1929, 77 per cent of the island's sugar was placed in the US market and made up more than one–half of that country's consumption. But the Depression cut US per capita consumption and the market was further restricted for Cuba by increased production from both protected domestic growers and those in the insular territories. By 1934, the Cuban share of a smaller market, at prices which had reached record lows, had fallen to one–quarter.[7] Exports to other markets fell less dramatically by volume,

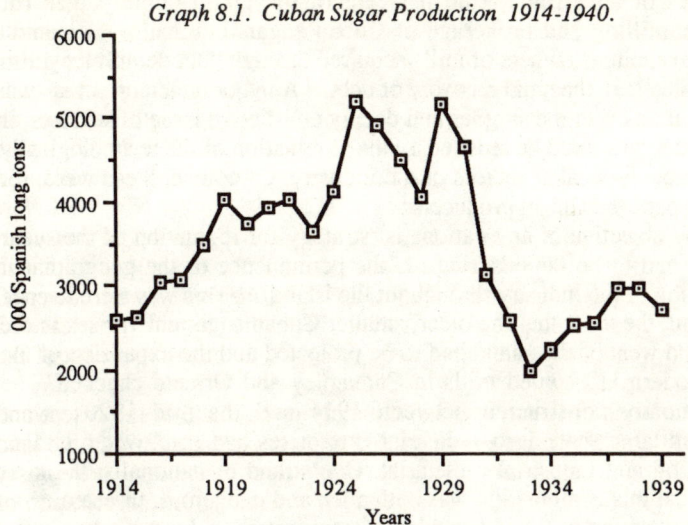

Graph 8.1. Cuban Sugar Production 1914-1940.

Source: Based on figures in, *Anuario Azucarero de Cuba*, 1952, 92–3, 95.

but prices remained extremely low. In consequence Cuban sugar production fell, from 5.2 million tons in 1929 to below 2 million tons in 1933.[8]

The collapse of Cuban exports entailed a comparable decline in imports and a steep fall in national income and employment. The ensuing social turmoil brought the overthrow, in 1933, of the President who had held power since 1925, but the government of nationalist complexion which succeeded him was short–lived. It was replaced in 1934 by a new political system whose central function was to regulate and control both the sugar industry and organized labour in the interests, respectively, of economic and political order.[9]

The key question, of course, was in whose interests was government to regulate the sugar industry? It was a question that had been posed most sharply during the crisis of 1925. In general, Cuban interests had been more enthusiastic in that year than their US counterparts about output restrictions to boost prices. Partly, this was because with mainland refineries, many US enterprises could still profit from having access to cheap raw sugar. Furthermore, the heavy overheads of the more modern large–scale US mills did not encourage them to seek output restrictions. It was this very fact which prompted Cuban producers to blame the US mills for creating the problem of over–production in the first place. The US banks were in an ambivalent position with regard to output limitation. As purveyors of an estimated 80 per cent of the credit employed in the growing, milling and brokerage of Cuban sugar they had a nationalist perspective, but as owners of mills acquired through debt default they also had an interest in the rapid recovery of debt.[10] A major undercurrent in such considerations was the simpler and deeply conflictive issue of whether or not production should be reduced by the elimination of the technologically and financially weaker sectors of the industry. Of course, these were, for the most part, the Cuban producers.

A key objective of any nationalist strategy for regulation of the sugar industry had to be the ensuring of 'the permanence of the geographical distribution of the industry throughout the island'.[11] This was a crude code expressing the view that the older, smaller Cuban sugar enterprises in the centre and west of the island had to be protected and the expansion of the more modern US owned mills in Camagüey and Oriente checked. The latter, mostly constructed between 1914 and the mid–1920s, were integrated large scale agro–industrial enterprises and employed a sizable agricultural and industrial proletariat. One strand of nationalist thinking argued that this combination was both alien and dangerous, threatening as it did the position of the independent cane growers (*colonos*), who, as the Cuban equivalent of a 'bold yeomanry' were viewed as a force for social stability in the countryside. These *colonos* were heavily concentrated (80 per cent) in the central and western provinces. The nationalists were given further cause for concern because the US mills imported large numbers of immigrant workers, 250,000 mostly seasonal labourers coming in from Haiti and Jamaica between 1913 and 1925.[12] It was grist for the Cuban nationalist mills that so much labour, as well as capital, in the giant US sugar enterprises of the two eastern provinces was alien. That this labour was also black tended to give a racist edge to this aspect of nationalist sentiment.

The more elemental mechanisms of the 'nationalist project' for the protection of Cuban sugar interests were first displayed with the passage of the Verdeja Act in 1926. This called for a 10 per cent reduction in output for each mill but stipulated that the cane supplies of *colonos* was not to be displaced by cane grown on lands directly administered by mills. While the act was limited in scope and lasted only two years it was a clear prototype of state regulation that implicitly favoured local interests over those of US capital.

But a renewal of this policy was not to come about until other attempts to hold back output and push up prices had been seen to fail. The most spectacular example and one which much inflamed nationalist feeling, was the outcome of the so–called 'Gentlemen's Agreement' of the Chadbourne Plan of 1930. This agreement sought to control supplies for the US market by restricting production in Cuba, the US, Puerto Rico and the Philippines. While Cuban output was drastically cut over 1931–33, that of her partners was increased and Cuba's share of the US market was halved. The Cuban delegation was led by Thomas Chadbourne, a New York corporation lawyer with direct interests in two mills and close links with the Chase National Bank.[13] The failure of his plan and of the agreement tended to discredit those advocating an identity of Cuban and US interests in both the national and international sugar economy and strengthened the organized pressure and political power of those emphasizing conflicts.

This conflictive view was given added weight because while the quotas allocated under the Jones–Costigan Act of 1934 tended to check the displacement of Cuban sugar from the US market, they were based on the average shares in that market between 1931 and 1933. This, of course, was precisely the period when there had been such spectacular breaches of the 'Gentlemen's Agreement'. Not surprisingly, it was a strongly–held Cuban opinion that the US Congress and Administration had thus unilaterally legitimated unfair conduct and institutionalized a process which had robbed Cuba of her place in the US market. It was clear that Cuba was unlikely to regain that lost position, but one firm purpose of Cuban producers in field and mills from 1934 was to ensure that there would be no further erosion in their share of production for that reduced market. In the event, they succeeded not only in checking but reversing the pre–Depression trend for foreign, primarily US, sugar interests to increase their share of total Cuban production.

The basic mechanisms to protect Cuban producers were those laid down in 1926. They could not be really effective, however, until after 1934. From 1932 to 1934, there were limited legislative attempts to protect smaller mills by permitting them to grind a quantity of cane closer to their proven capacity level than was allowed for high capacity mills.[14] Nonetheless, with US prices so low such protection could do little to alleviate the general crisis of profitability which affected both Cuban and US interests. This was exacerbated by a restriction in total production yet greater than the fall in the volume of exports in order that heavy stocks held over from the crops of 1929 and 1930 could be shed on non–US markets over a four year period. All this changed after 1934.

Up until 1933, the differences between London and New York sugar prices were relatively unimportant. However, from this time, and especially after the allocation of the 1934 quotas, prices for the US and London (world) markets began radically to diverge. Accompanying the improvement of US prices for raw sugar were two tariff reductions pushed through by the Roosevelt Administration in 1934, which cut the rate from 2 cents per pound to 0.9 cents. This did not affect the restricted access of Cuban–produced sugar to the US market, but it did increase the profitability of producers enabled to recover a larger portion of the mainland sale price.

These US tariff changes were accompanied by modifications in the Cuban tariff structure, designed to offer comprehensive protection for US exporters of manufactures and foodstuffs from both other foreign as well as domestic competition. Together with the 'good neighbourly' abrogation of the hated Platt Amendment of 1902, this package formed the heart of the New Deal strategy for US–Cuban economic relations, the goal of which was to boost the fortunes of US investors through economic regeneration and political stability.[15] The counter–strategy of Cuban interests in the sugar industry was to ensure that their fortunes were also salvaged. This was to be done, at least partly, at the expense of US and other foreign sugar interests.

The formal implementation of the nationalist project was in the hands of the Cuban Institute for Sugar Stabilization (ICEA). This allocated total output for all home and overseas markets between the mills. Each mill's quota comprised a quantity of sugar that was weighted in proportion to total national production for each market and was fixed at their respective, and quite different, price levels. Thus mills producing solely for the local market or for the world market in fact received prices boosted by the relative share of national production destined for the more higher–priced US market. The effect of this was the erosion and redistribution of the more remunerative income–advantages that would otherwise have accrued to those mills, largely US owned, with close ties to the mainland market. In this way the smaller, weaker Cuban mills were able to capture for themselves at least some of the higher rates of profit derived from export to the US.

The other major issue on the nationalist agenda was the fate of the *colonos*. Their position became increasingly precarious as the crisis deepened, particularly in the provinces of Camagüey and Oriente, where, according to one report, as much as 90 per cent of the small and medium *colonos* were displaced between 1930 and 1934.[16] The anxieties of this numerous class of farmers were reflected in the nature of their demands and in the organized vigour with which they were pressed in the thirties.

The *colonos* and the mill owners were represented in the ICEA, which was initially established in 1931 to administer the Chadbourne Plan. There were eighteen members, of which six represented the larger mills, six the smaller ones, and six represented the *colonos*. This strengthened the voice of Cuban interests. Moreover, the cane growers, organized into the Cuban Cane Growers Association, greatly benefited from the divisions and infighting between the representatives of the small, mainly Cuban mills and the large, predominantly US mills. Both inside and outside the ICEA, the *colonos* could also take full advantage of pervasive nationalist sentiments by stressing their historic, and geographically concentrated, association with the 'Cuban' sugar economy of the central and western provinces as contrasted with the 'alien' plantation system in the east. Furthermore, the majority of *colonos* did not use wage–labour. This ensured that their organized activities and demands complemented rather than conflicted with the simultaneous mobilizations of rural workers to improve their conditions in the fields and mills. Since the primary targets of both most *colonos* and most rural proletarians were the mills and their plantations, it is not easy to disentangle the practical impact of their respective actions. The issue was further confused in that the activities of social groups pressing different claims against interlocking agricultural and industrial sugar enterprises could plausibly, if not entirely accurately, be portrayed as nationalist struggles against foreign or imperialist interests.[17]

In the extreme but common situation of actual or threatened mill–closures, the most urgent interests of the *colonos* were inseparable from those of the broader rural community. Between 1929 and 1933 no less than 38 mills were closed, most being Cuban owned and in the central and western provinces. Such closures did not simply eliminate the employment and income of direct employees and suppliers. The livelihood of entire rural communities and provincial towns could depend directly or indirectly on the mills nearby. Because of this, 32 of the mills were reactivated between 1933 and 1937[18], and the protective legislation facilitating this was enacted and enforced in the context of turbulent community struggles that transcended the pursuit of narrow sectional interests.

The principal objectives sought and, by 1937, in large part secured by the *colonos* were: guaranteed participation in the share of cane ground by mills, biased in favour of *colonos* in general and with minimum quotas for the smallest *colonos* in particular; security of tenure; low rent ceilings with a moratorium on the repayment of debts incurred during the worst years of the Depression; and a guaranteed proportion of the price of raw sugar received by the mills. Some of these measures were partially applied before 1937, but they were consolidated and strengthened in that year in the Sugar Coordination Law. With the implementation of this law the *colonos* arguably emerged as '*los grandes triunfadores*' of the revolution of 1933.[19] In many ways this result was not surprising for the *colonato* embraced large, medium and small–scale Cuban cane growers and enjoyed the support of an extremely broad spectrum of the contending factions in national political life.

The rural workers were less well placed, but by comparison with its desperate situation in the early 1930s, labour in the sugar economy also gained from the events of 1933 and their aftermath in terms of trade union legislation, social security provisions and improved wages and working hours. Its annual level of employment and wage–income throughout the 1930s remained below that associated with periods of higher production and prices in the 1920s but the mill–quota system which rationed output also restored or conserved a spatially diffused rationed employment of mill and field labour. For the labour force of the sugar industry as a whole there was a reduction in total days worked, but this was preferable to the alternative of a maintained level of employment for some and total unemployment for others.

Such changes were not welcomed by the US sugar interests in Cuba. Between 1934 and 1937 they were faced with increasing institutionalized frustrations. Regulation restricting production and protecting smaller mills prevented the US mills from realizing the economies of scale associated with their higher capacity operations. Quotas based on differential pricing made it impossible for the foreign enterprises to profit fully from their vertical integration with mainland refineries. They were also being disadvantaged by restrictions on the import of cheap workers and by the growing strength of organized labour. In agricultural operations, the preferential allocation of cane quotas to *colonos* at the expense of mill administered cane first checked and then reversed the development of integrated, agro–industrial plantations. Rents fixed at comparatively low levels, security of tenure for *colonos,* and higher prices to be paid for *colono* cane in general all reduced the rate of return to be expected from the enormous areas of cane land bought, or acquired by default, by US banks and sugar companies prior to the 1930s.

The impact of these policies was soon apparent. In the period 1929–1933 the number of active non–US mills fell sharply by comparison with their US–owned counterparts, but after 1934 their reactivation was also greater. There were six fewer US mills grinding in 1939 than in 1929, but the number of non–US enterprises was the same as at the onset of the Depression. This change was reflected in production data which show that in 1934 US mills ground 64.1 per cent of national production but in 1938 only 58.1 per cent.[20] Over the year 1934 to 1939 there was also a significant erosion in the share of national production coming from the US dominated eastern provinces in favour of the central and western regions where *colono* production was most heavily concentrated.[21]

Given these changes it was not surprising to find that from 1939 there was the beginning of a major repatriation of capital within the Cuban sugar industry. Prior to 1934 – and most notably between 1921 and 1926 – Cuban mills were being acquired in large numbers by foreign banks and sugar interests. After 1934, by contrast, there was no report of any Cuban mill passing into foreign ownership. Secondly, while there were significant intra–US and intra–Cuban exchanges of mill ownership between 1934 and 1951, the outstanding feature of the data reporting all mill sales was the transition into Cuban hands of a total of 47 foreign owned mills. Of these, 32 were reported to have been US–owned and nine Canadian.[22] As has been shown, between 1934 and 1939 the recapture of a significant share of production by Cuban nationals was promoted by protective state regulations encouraging the resumption of activity of numerous small–scale mills that had shut down in the worst years of the Great Depression. From 1940, however, the yet more important increase in the Cuban share of production was associated primarily with the repatriation of an important part of the sugar industry itself.

Too great an emphasis on the recovery of mills previously in foreign hands could obscure the dual mechanism whereby formal or *de facto* Cuban ownership or control of once alienated lands were also extended over these years. An obvious expansion of such ownership occurred with the purchase of foreign mills and their lands. A less formal method followed from the Sugar Coordination Act of 1937. The security of tenure it provided, together with the institutionalized discrimination against mill–administered cane, led to the extension of the *colono* system. In the 1920s, perhaps 70 per cent of the cane was grown by *colonos* whereas by 1950 this had increased to at least 85 per cent and probably more. This increase was primarily due to the leasing to *colonos* of mill owned lands. Ordinarily one would not point to a proliferation of tenant farmers as constituting any significant erosion of the power of landlords. However, it was notable that while foreign sugar interests continued to own a far greater proportion of cane land than was comprised by their own plantations, the bulk of these holdings was concentrated virtually inalienably in the hands of a relatively small number of wealthy, politically powerful Cuban *colonos* whose formal designation as tenants would in other agrarian contexts seem absurd.

Conclusions

Given the salient features of the Cuban sugar industry in the late 1920s, few would have predicted the course of events which was actually to unfold. The most plausible candidates for survival, as the years 1930–33 continued to demonstrate, were obviously the financially stronger and technologically more advanced US agro–industrial complexes. It was, of course, effective and partisan state intervention and control that frustrated the continued operation of the law of the survival of the fittest. While criticized as inhibiting the free operation of market forces, such actions reflected the comprehensive negation by successive US administrations of all free market principles determining access to the US sugar market. Protective domestic market controls imposed by the US legitimated corresponding protective production controls imposed by the Cuban Government. Such general political legitimation, however, could not itself explain the successful, partisan political actions which defended Cuban sugar interests at the expense of foreign ones. This required an exercise of effective state political power reflecting attributes of national sovereignty notably absent in earlier periods.

Two major standard works on the Cuban economy, Ramiro Guerra y Sanchez's *Azúcar y Población en las Antillas* (1927) and Leland Jenks's *Our Cuban Colony* (1928), expounded a view of twentieth century economic history in which US sugar investment in the island had grown to be the decisive force in US investments as a whole and in which the political power of the US state reinforced the formidable power of US private corporations. The combination yielded an image of the US sugar interest as a kind of politico–economic steam–roller that inexorably flattened the puny obstacles that competing Cuban interest might seek to place in its path. Such an impression was difficult to challenge convincingly in the 1920s, and it rendered politically as well as economically implausible the erosion of the dominance of the US sugar interests from 1934. This required not simply a politically potent agglutination of Cuban social forces in a struggle to defend the national sugar economy. It also demanded a corresponding inability of US sugar interests to marshal political forces that were sufficiently powerful to defeat them. This relative political weakness reflected the fact that in times of falling US sugar prices, the decisive battles for political influence with the US Congress and Administration were not fought between US sugar interests in the island and their weaker Cuban counterparts. Instead they were fought between representatives of the Cuban sugar economy, spearheaded by US interests within it, and rival US continental producers supported by those of the colonial territories of Puerto Rico, Hawaii, and the Philippines.

Up to 1934, the favoured weapon of US protectionism was the tariff. In the brawl for political leverage in Washington's sugar lobbies, the preeminence of US interests in Cuba's sugar exports failed, most crucially in 1930, to render these more 'American' and less 'foreign' in origin. On the contrary, the tariff increase displacing Cuban sugar from US markets from 1930–34 castigated Cuba its for juridical status as an independent sovereign state while it rewarded formal colonies with the status of protected domestic producers. As Cuban exports, imports and employment collapsed in consequence, advocacy of economic advantages of *de jure* national independence (circumscribed by the Platt Amendment) but *de facto* national subordination to US sugar interests lost all plausibility.

The Revolution of 1933, symptomatic of the gravity of Cuba's political and economic crisis, intensified the US Administration's concern for social order in the island. But, the US was also concerned about similar problems in her direct island dependencies, and, here, the preoccupation with Cuba was not to be translated into the restoration of the Cuban–US sugar trade to its pre–1930 levels. Between 1929 and 1934, Cuba's share of a declining US sugar market was halved. This was a disaster for the sugar economy, whether domestically or US owned. Viewed from Washington, however, the very enormity of Cuba's resulting production problem solved the difficulties facing US mainland growers and those in Hawaii, Puerto Rico and the Philippines. In contrast to the relatively depressed level of Cuban production during the 1930s, output in these islands was maintained or increased.

Such a consolidation of the gains of Cuba's rivals for access to the US market entailed a protracted exacerbation of social tensions in the island requiring alleviation by other means. At the level of formal inter–state relations, a conspicuously combustible element fuelling Cuban nationalist passions was removed in 1934 when the Platt Amendment was abrogated. The crisis of profitability for both national and US–owned sugar enterprises was ameliorated by improved US sugar prices and by sugar tariff cuts. The latter were conditioned by reciprocal cuts in Cuban import tariffs that reinforced the access of US commercial interests to non–sugar sectors of the economy. This is turn further retarded the development of incipient Cuban production in such sectors. There was, however, important compensation for the coalition of Cuban sugar interests. From 1934, powerful and indentifiably Cuban interests in large and small–scale growing and in sugar manufacture effectively linked the sugar economy to the political state and, within limits, imposed their own authority upon the state and the institutional machinery apportioning the costs of the protracted depression in production. From 1934, the consolidation and implementation of a formula originally devised in the mid–twenties first checked and then reversed the secular twentieth century decline in the productive scope of Cuban interests in the sugar economy at the expense primarily of US sugar interests.

References

1. Manuel Moreno Fraginals, *El Ingenio*, Vol. III, Havana, 1978, Table 1, 39.

2. *Anuario Azucarero de Cuba*, Havana, 1952, 92–3, 95.

3. Leland H. Jenks, *Nuestra Colonia de Cuba*, Buenos Aires, 1961, 49.

4. Hugh Thomas, *Cuba, or the Pursuit of Freedom*, London, 1971, 553.

5. Jenks, *Colonia*, 261; Lionel Soto Prieto, *La Revolución del 33*, Vol. I, Havana, 1977, 261.

6. J.R. Benjamin, *The United States and Cuba: Hegemony and Dependent Development, 1880–1934*, Pittsburgh, 1977, 203.

7. *Anuario Azucarero*, 199.

8. *Ibid.*, 92–3, 95.

9. Jorge Dominguez, *Cuba: Order and Revolution*, London, 1978, 84.

10. Benjamin, *Dependent Development*, 29–33.

11. Ramiro Guerra y Sanchez, *La Industria Azucarera de Cuba*, Havana, 1941, 263.

12. R. Guerra y Sanchez, *Azúcar y Población en las Antillas*, 6th ed, Havana, 1961, 187.

13. Benjamin, *Dependent Development*, 34–5, Soto, *Revolución*, Vol.II, 274.

14. Guerra y Sanchez, *La Industria*, 50–2.

15. Benjamin, *Dependent Development*, 174, 177.

16. Soto, *Revolución*, Vol.II, 286.

17. Juan Martínez–Alier, *Cuba: Economía y Sociedad*, Paris, 1972, 75–108.

18. *Anuario Azucarero*, 101–2.

19. Martínez–Alier, *Cuba*, 80.

20. *Anuario Azucarero*, 101–2.

21. *Ibid.*, 92–3, 95.

22. *Ibid.*, 72.

9

The Uneasy Relationship: Peasants, Plantocrats and the Trinidad Sugar Industry, 1919–1938

Kusha Haraksingh

The problem

This paper seeks to explore the main response of the Trinidad sugar industry to declining fortunes in the interwar period. This was the further encouragement given to private cane farmers to cultivate canes for milling at the central factories owned and operated by the large sugar planters. Whereas in the twenty years preceding the end of WWI cane farmers in Trinidad contributed on average 33.9 per cent of the total crop processed at various mills in the island, in the subsequent two decades their share grew to 45.3 per cent. The more pronounced entrenchment of the cane farming sector was not without its ups and downs, and at various stages many farmers had ample cause to question their continued participation in cane farming as a remunerative economic activity. The total domination of the industry by the sugar manufacturers, based on the passage of a series of legislative measures by the Colonial Government, created serious problems. Low and fluctuating returns likewise generated a great deal of anxiety which infrequent years of good prices did little to dispel. Nevertheless, the number of people classified as cane farmers hardly declined at all; furthermore, with the natural growth in the population the amount of people who relied on cane farming – either as farmers themselves, or as dependents or employees of farmers – for at least part of their livelihood continued to increase. This raises the question of the overall relationship between returns to farmers, as determined by the sugar plantocracy, and the growth of the cane farming sector, and illustrates the operation of the strategy adopted by the Trinidad sugar economy to cope with the changing international market of the interwar period.

The background

The people referred to as cane farmers cannot all be described as peasants in the orthodox sense of that term. They were a highly stratified group encompassing at the bottom, people who owned or rented relatively small parcels of land which they cultivated mainly with family labour, and at the top, farmers with considerable estates providing employment for a significant labour force.[1]

Table 9.1 records the number of cane farmers for the period 1919–1938. Following WWI the number of farmers climbed to a peak of 26,425 in 1921, then fluctuated around the mean for the entire period of 19,426 until a trough represented by 15,102 farmers was reached in 1938. On the surface it would appear that the war years' boom in sugar prices accounts for the 1921 peak, and that the lean fortunes of cocoa on the world market in the mid–1920s explain the increase in cane farmers in 1926. But it is problematic to carry any analysis based on the absolute numbers of cane farmers too far since the data suffer from several defects.

In the first place each estate was in the habit of opening an account for every farmer from whom canes were purchased so that a farmer supplying canes to more than one estate would be counted more than once. The reverse of this also applied. When a middleman or contractor supplied canes to an estate on behalf of several farmers, they were all considered one supplier and, for the purpose of Table 9.1, one farmer. In addition, some farmers occasionally resorted to selling their canes under various aliases as a means of evading repayment on loans which they had secured from the estates. Several sources indicate that this practice was widespread following the relatively low prices which farmers obtained for their canes in 1921 and 1922.[2] Finally the low figure recorded for 1938 arose not from an absolute decline in the number of farmers but from the implementation of a new system of classification.

Thus, the only safe conclusion would be that the number of cane farmers remained fairly constant throughout the period – but more so the number of middling and large farmers than the number of small farmers. Table 9.2 gives a profile of the farmers according to tons supplied for the closing years of the period when data are available. It appears that the greatest seasonal variation occurred among the small farmers, that is, those supplying less than 20 tons of cane. This group also contained the highest proportion of cultivating lessees. As a matter of fact, if farmers are classified according to the nature of their land tenure then as we go from a lower to a higher tonnage class the ratio of owners to renters also increases. Where very large farmers are concerned, that is, those producing more than 1,000 tons of cane, there are indeed more owners than renters.

The small farmers, who according to the data for 1936 and 1937, comprised about 65 per cent of the classified suppliers, supplemented their

Table 9.1. Number of Cane Farmers.

Year	Number	Year	Number
1919	20935	1929	18081
1920	25360	1930	18285
1921	26425	1931	17978
1922	21350	1932	17440
1923	16500	1933	17852
1924	17068	1934	15457
1925	19474	1935	18062
1926	22939	1936	19471
1927	19385	1937	20705
1928	20668	1938	15102

Source: *Report of the Trinidad Island–wide Cane Farmers Association*, various years.

Table 9.2. Cane Farmers by Tons Supplied.

	1936	1937
Under 5 tons	4517	5003
6 to 20 tons	6600	7474
21 to 50 tons	3970	4498
51 to 100 tons	1384	1533
101 to 500 tons	643	724
501 to 1000 tons	45	55
Over 1000 tons	36	37
Unclassified	2276	1381
Total	19471	20705

Source: *Trinidad Legislative Council Paper 119*, 1940. Cane Farming Committee Report.

income from cane farming by working on nearby estates. In addition, they had worked out an ingenious scheme of intercropping on the cane lands which enabled them to produce food crops without hampering the cultivation of the canes.[3] As the *Ste. Madeline Quarterly Report*, a journal published by Ste. Madeline Sugar Estates, said in 1922, intercropping was the secret of the farmer's success. There can be no doubt that the farmer's ability to make effective use of the land was a crucial factor in determining the level of his existence. But intercropping came to be used to rationalize low prices for canes. The argument was that the farmers were able to absorb the effect of poor cane prices since they did not depend entirely on cane farming for their livelihood. It was not the first time that the peasant's industry was used in argument against him.

In 1933, a Committee set up to report on cane farming highlighted the industriousness of the cane farmer and his family as 'undoubtedly the chief factor leading to success'.[4] The work involved in cane farming included a series of arduous activities – planting, cultivation and harvesting – which were executed with various kinds of hand implements. First of all, areas to be planted would have to be cutlassed, cleared and drained and the plots divided into rows by banks. By the mid–1930s the cost of preparing the land for planting could amount to 20 Trinidad dollars ($4.80=£) an acre but in fact this work was performed by the farmer and his family who almost never reckoned the cost of their own labour. The fields would normally be planted in March–April with cuttings drawn from the 'tops' of growing canes. Some but not much planting also took place in October–November. On the smaller farms planting would involve not entire plots of land but simply the placing of cuttings in the ground where adequate spaces between the growing canes could be found. The operations associated with planting – cutting tops, dropping the cuttings into holes, and breaking the banks – were estimated to cost up to 10 dollars per acre in 1933, and again were performed mainly by the peasant and his family. Domestic labour was also involved in the forking, weeding, and trashing which took place in the first few months of growth. It is estimated that in money terms the careful farmer, through the use of family labour, would have saved about 20 dollars per acre on these activities. During the early growing period, too, the farmers produced food crops along the banks. From 12 to 18 months after planting the cane was ready for harvesting. This would mean that canes planted in the spring or autumn of 1920, say, would not normally be delivered to the factories until the crop season of 1922. During the harvesting period, from January to June, some extra–family labour might be engaged for cutting and carting canes to the scales. The rate in the 1930s for cutting one task – that is, an area that yielded about two to three tons of cane – was 50 cents, and carting charges were one dollar per ton. Congestion at the scales restricted the farmer's delivery to one cart–load of cane a day and the advent of the rains, normally in late June, signalled the end of the harvest period by rendering the tracks to the fields impassable.

Continuous figures of wage rates are not available, but there is enough information to conclude that even with the maximum input of family labour, the cultivation and harvesting of plant canes was for the small farmer an expensive undertaking. The 1933 Report indicated that the proceeds of the sale of plant canes barely covered the cost of producing them and that the farmers had to rely on ratoons for any appreciable margin of profit. Ratoon cane, cane produced from the stocks of an earlier harvest, was ready for reaping in 12 months but each year there was a progressive decline in the yield per acre. For example, while on the sugar plantations and on well attended farmers' fields plant canes could produce up to 40 tons per acre, a first ratoon would yield 30 tons and a second only 25 tons per acre. The small farmer, according to the Chairman of the Ste. Madeline Sugar Company, could average only half the weight per acre which the estates achieved, and he would normally continue with an annual harvesting until the yield per acre reached an unacceptable level.

There is general agreement about the crucial role played by the plantocracy in Trinidad in fostering the various activities which make up cane farming. There is also consensus on the motives of the planters in encouraging the cane farming industry.[5] Generally, two overriding reasons are cited. First, the planters hoped to encourage settlement near their estates with a view to ensuring for themselves an adequate labour supply. A similar consideration in British Guiana led to the promotion of rice cultivation on lands adjacent to the sugar estates. Secondly, the planters hoped to reduce their expenses through their access to farmers' canes which they expected to procure at prices lower than the cost of producing their own canes. Thus, some planters urged the opening up of crown lands adjacent to their estates to potential farmers and also made their own estate lands available. Some individual planters are credited with the most direct encouragement; one in fact, turned over his entire estate to cane farmers, and quite a few were extremely vocal in supporting the expansion of the railway system in the countryside as a means of developing the cane farming sector. The planters also provided loans, hoping thereby to free the would–be cane farmer from dependence on other sources of credit such as the private money lender or village shopkeeper but simultaneously to bind him to the estates. The stimulus provided by the planters was buttressed by other factors. These included attitudes towards wage labour, the desire for independence, considerations of status and the pressure of a growing population. In all of this material returns would strike one as a significant factor.

For our purposes here, we want to isolate the factor of material returns as a 'predictor' of cane farming and to test the responsiveness of the cane farmer to the movement of prices. We may begin with the straightforward assumption that farmers do react to the prices commanded by their produce. The expectation of high prices normally encourages them to expand production; similarly, the expectation of low prices induces them to curtail their production or to shift to more lucrative crops. There are scattered pieces of qualitative evidence that lead us to suppose that cane (and cocoa) farmers in Trinidad behaved in this way. For example, the Trinidad Sugar Estates report for 15 months ending September 1923 pointed out that the company had produced less sugar than was expected mainly because low prices for farmers' canes in 1921 and 1922 had caused farmers to reduce their planting. This was confirmed by the Director of Agriculture who in his report for 1923 indicated that the fall in the price of sugar had affected cultivation of that year's crop. Two years later, the Director was reporting that farmers were responding to unfavourable cocoa prices by shifting to cane cultivation. What needs to be examined, more systematically than the qualitative evidence alone will allow us, is the strength of the relationship between the prices secured by cane farmers and the level of their output. The first task therefore is to establish what happened to cane production in the period between the wars.

Output

The data with which we have to work to establish trends in cane production are tons of cane processed at the mills, broken down according to source of supply (Table 9.3). This is not a perfect but is certainly a tolerable replacement for tons of cane produced, for which figures do not exist. There is the theoretical possibility that tons processed may not accurately reflect tons grown if limited milling capacity, for instance, forced the sugar manufacturers to restrict their purchase of farmers' canes; in practice this did not apply in the years 1919–1938. It is only towards the close of this period that the limitation of farmers' cane became a real issue. However, interviews with erstwhile cane–weighers, whose task it was to relay information from the weigh–bridges or scales at the purchasing points to the estate office where the returns were compiled, have indicated that some amount of liberty was taken with the figures. But there appears to have been no systematic bias or error. More importantly, a significant discrepancy between canes processed and real production could have arisen when early rains curtailed the harvest. Canes were then left standing over to be harvested in the subsequent year's crop, as was the case in 1929 and again in 1932.

The ideal alternative to annual data on tons of cane produced by the farmers would have been data on acreage under farmer's cane. This has certain obvious advantages when compared to tons processed. For one, the latter information may mislead us about the planting activities of farmers in years when the size of the harvest was affected by drought, fire or froghopper attacks. However, though it is possible to construct a continuous series relating to cane acreage from the Blue Books, the figures reported there are not very precise and in any case make no distinction between lands under estate and lands under farmers' cane.

Nothing is to be lost by regarding the data on tons of canes processed as roughly equivalent to the size of the harvest. Bearing this in mind, one may take it from Table 9.3 that between 1919 and 1938, total cane production increased from 546,000 to 1,292,000 tons. The turning point was the year 1932 when for the first time cane production exceeded one million tons. More importantly the figures show that notwithstanding their encouragement to cane farmers, the planters were increasing their own production at a rate more than one and a half times faster than that of cane farmers. Whereas farmers crossed the half million mark in 1936, the estates had done so fully five years earlier. Not all of this more rapid growth on the part of the estates had to do with expanding acreage. Some of it could be explained by the introduction of improved methods of cultivation, including shorter periods of ratooning and the more liberal application of fertilizers – as compared both to earlier years and to farmers' plots. In the case of the cane farmers as a whole, however, and especially with respect to those supplying less than 200 tons of cane per annum, increasing production resulted primarily from the widening of the area under cane and only marginally from changes in cultivation methods and practices.

Where farmers' output is concerned the years with the largest deviation from the trend are 1923 and 1937. The earlier fluctuation was associated with a fall in the returns per ton to farmers from \$4.84 in 1921 to \$2.57 in 1922 – that is, a drop of almost 50 per cent. The later fluctuation was connected to the poor prospects of cocoa on the world market which caused some farmers to shift over to cane cultivation. In other years the movement away from the trend can be explained to some extent by a number of factors including froghopper attacks in 1919, 1924 and 1926, early rains in 1927, 1929 and 1932 and drought in 1921 and 1934.[6] All of these influences together with the price paid to farmers for their canes comprise the irregular short term factors which helped to determine cane output.

Returns and the effect on output

One of the most troublesome questions with which the sugar industry has had to contend, and indeed one that is still the source of much dissension and agitation, is the method of determining the price to be paid to farmers for their canes. In the early days of cane farming in Trinidad farmers were remunerated according to a flat rate system. The farmers considered this formula unsatisfactory since it did not allow them to share in the increased profits which the millers realized in those years when sugar fetched favourable prices. In other years the millers for their part decided that a fixed price was not good business. The general dissatisfaction with the flat rate system led eventually to the introduction by some millers in 1907 of payment according to a sliding scale. Thereafter, a number of such scales were put into operation, including a Government recommended scale in 1916, a scale devised by the Ste. Madeline Sugar Company in 1919, which was adopted by most of the companies in 1924–25, and a legally enacted sliding scale in 1930.

Table 9.3. Tons of Cane Processed by Source of Supply.

Year	Farmers(nearest 500 tons)	Estate(nearest 500 tons)	Total
1919	270,500	275,500	546,000
1920	344,000	319,500	663,500
1921	389,500	287,000	676,500
1922	355,500	340,500	696,000
1923	186,500	253,500	440,000
1924	237,500	324,500	562,000
1925	327,500	395,000	722,500
1926	359,000	384,000	743,000
1927	286,500	330,500	617,000
1928	375,000	449,500	824,500
1929	400,000	426,500	826,500
1930	303,500	458,000	761,500
1931	385,000	559,500	944,500
1932	455,000	577,000	1,032,000
1933	488,000	666,000	1,154,000
1934	370,000	557,000	927,000
1935	402,500	603,500	1,006,000
1936	594,000	780,000	1,374,000
1937	630,000	813,000	1,443,000
1938	571,000	721,000	1,292,000

Source: *Report of the Commission Appointed to Enquire into the Working of the Sugar Industry of Trinidad*, Port of Spain, 1949.

The scales in operation between 1919 and 1938 had several features in common. Firstly, the price to be paid to farmers was linked to the price of sugar and in particular to the average f.o.b. price of grey crystal sugar at Port of Spain from 1 January to 30 June. Secondly, provision was made for deductions to cover the cost of transport and bags and to defray government taxes and selling charges. Thirdly, the sum remaining after the deduction of charges was split between farmer and manufacturer according to a fixed percentage – 4.5 per cent to the farmer in 1919 and 5 per cent in 1930. Fourthly, the system of remuneration included the concept of an initial payment soon after the delivery of canes, and a deferred payment or 'back–pay' consequent on the declaration of the f.o.b. price of sugar. In any given year, then, the farmers would not know the total price before the end of June by which time the harvest period was terminated. Finally, the sliding scales took no account of the profits to be derived from the sale of by–products such as rum and molasses.

It is clear that in the sliding scale the millers had found a system of payment which ensured that the effect of any falling off in the sugar price would be passed on to the cane farmer. The sugar price and the price paid for farmers' canes run in parallel directions, revealing a correlation coefficient of 0.95, an almost perfect positive association between these variables. This picture is quite different when the relationship between farmers' price and output is investigated. The variables which were employed in this investigation are listed in Table 9.4 and in Table 9.5 the correlation and regression coefficients which were produced are listed. In all the calculations output is lagged two years behind price since that is the time phase between planting and harvesting. In other words, a distance of two years separates the price at harvest time from the price at the time of planting.

The first correlation (r14) was performed on the raw data values of output and price with a two–year lag. The results indicated that the price for farmers' cane obtaining two years before the harvest hardly bore any relation to the size of the harvest. Put another way, this means that the raw price did not have any significant effect on planting. Raw price in year T1 'explained' only 10 per cent of the variation in harvest in T3.

To remove any suspicion that the long–term trend in output was distorting the results a bivariate regression of detrended output on price (r24) was performed, again with a two–year lag. The results were even more insignificant than for r14. A similar calculation involving detrended output and detrended price (r29) – in the event that the price trend was obscuring the picture – produced slightly more encouraging results. The next step was to look at changes from year to year by taking first differences of prices and of output. In a series where there is considerable upward and downward fluctuations around a trend, this manoeuvre emphasizes those fluctuations and removes the importance of the trend. The correlation of first differences output and first differences price (r35 with the usual lag) produced a coefficient of 0.30 suggesting a slight link between annual changes in price and annual changes in the size of the harvest.

The results of r35 raised the possibility that the price trend itself helped to determine planting and therefore output. There is some theoretical support for this contention. Since it is the price at harvest time which regulates the farmer's income he must, if rational, project some price two years ahead when he decided what acreage to plant. He could only project this trend on the basis of what had been happening to prices for some years before planting.

To test whether the price trend was having any effect on planting activities simple regressions were run involving a two–year, three–year and four–year moving average of price as independent variables and output as the dependent variable (r16, r17 and r18). The results give some empirical credibility to the argument that what had been happening to prices for two years before planting, and even more significantly, for three years before planting, carried some weight with the farmer. The correlation of a three–year moving average of price and output was –0.79 indicating that

117

Table 9.4. List of Variables.

Variable 1	Output of Farmers' Cane.
2	Detrended Output.
3	First Differences Output.
4	Farmers' Cane Price.
5	First Differences Farmers' Cane Price.
6	Two–Year Moving Average Price.
7	Three–Year Moving Average Price.
8	Four–Year Moving Average Price.
9	Detrended Price.

Table 9.5. Correlations.

	R	r2
r_{14}	−0.320	0.100
r_{24}	0.185	0.034
r_{29}	0.350	0.130
r_{35}	0.308	0.095
r_{16}	−0.590	0.350
r_{17}	−0.790	0.620
r_{18}	−0.510	0.260

when prices had been falling for three years, the farmers increased planting in the hope that at harvest time prices would be on the upturn. The three–year moving average of price explained 62 per cent of the variation in output. A longer term movement of prices had less explanatory power.

Conclusion

A witness who claimed to have been a cane farmer for twenty years told the 1926 Commission that 'the best encouragement which you can give a farmer ... is good prices for his canes'.[7] That might indeed have been so, but in the years between 1919 and 1938 the overall downward trend in returns to farmers was associated with an increase in their output of cane. This raises the possibility that despite the direction of prices, the returns which were realized were in fact considered as favourable by the farmers. However, this can be discounted in view of deep and continuous discontent which farmers expressed over cane prices. A more plausible explanation might be the deep hunger for land which could partly be satiated by entering into a cane farming contract with a sugar plantation. This would oblige the farmer to plant and harvest cane or risk the termination of the arrangement, but once he had access to the land he could use it for his own purposes. For the most part he would, whenever possible, grow food crops both for his domestic consumption and for sale in the local markets. Nor must considerations of status be discounted. In the sugar villages of Trinidad between the wars, the large farmer was top of the scale, the landless labourer at the bottom.[8] The labourer with access to land could comfort himself that he was not on the lowest rung.

A further explanation would be that the farmers did not expect that the price at harvest time would be the same as the price before planting. They operated on the hope that if prices had been falling for some years the trend would have been reversed by the time of reaping. But a crop like cane committed the farmer for a number of years and deprived him of the flexibility necessary to ride the upward and downward fluctuations in prices. Thus, if the cane farmer's expectations were not fulfilled he had little choice – indeed, there was nothing to be lost – but to go on reaping his ratoons. At the very least, he had a back–pay coming his way, and when that amount was eventually declared, he regarded it – since it came at the end of the crop when all his work for that harvest was done – not so much as part of his just reward but as a windfall, almost as interest on a saving account.[9]

The explanation that seems most sound, however, is that as the returns per ton of cane fell, cane farmers found it necessary to undertake more planting in order to maintain the level of their remuneration. The nature of the situation here was well understood by the sugar plantocracy who had used a similar strategy with respect to the labourers on their estates since the closing decades of the nineteenth century. Indeed, it was almost a fixed piece of planter wisdom that one way to ensure the performance of more tasks or units of work per labourer and hence from their point of view greater efficiency was to offer a lower rate for each task while at the same time using political influence and economic leverage to cut off alternative labour opportunities.[10] The big cane farmer when faced with falling returns would pass on some of his troubles to the workers whom he employed; it is well known that conditions of work on these private cane farms were worse than on the large commercial estates. But the small farmers, who comprised the bulk of the cane farming sector, had no such room for manoeuvre. The planters would ease the binds from time to time by increasing the price paid above the minimum provided for by the sliding scale. At the same time, however, they strengthened their domination over the cane farmers by the provision of loans, by the implementation of a contract system, and by undermining the security of tenure of those cane farmers who rented estate lands and were slackening in output. The downward trend in prices, together with these non–price factors, conditioned the farmer's view of his world, and set the parameters of his uneasy relationship with the sugar plantocracy.

References

1. British Parl. Papers, *Cmd 3517*, Report of West Indies Sugar Commission 1930.
2. *Trinidad Legislative Council Paper 70*, 1926. Report of the Cane Farming Commission.
3. *Trinidad Legislative Council Paper 119*, 1940. Cane Farming Committee Report.
4. *Trinidad Legislative Council Paper 84*, 1933. Report on Cane Farming.
5. H. Johnson, 'The Origins and Early Development of Cane Farming in Trinidad 1882–1906', *Journal of Caribbean History*, vol 5, Nov. 1972, 46–47.
6. *Trinidad Legislative Council Paper 1*, 1943. Benham Committee Report.
7. *Trinidad Legislative Council Paper 70*, 1920.
8. A fuller discussion is provided in K. Singh's 'Economy and Polity in Trinidad 1917–38', unpublished PhD thesis, University of the West Indies, Trinidad, 1975.
9. For a more comprehensive treatment see D. Maharaj, 'Cane Farming in the Trinidad Sugar Industry', unpublished PhD thesis, University of Edinburgh, 1966.
10. Argued in my article, 'Labour, Technology and the Sugar Estates in Trinidad, 1879–1914' in *Crisis and Change in the International Sugar Economy*, B. Albert and A. Graves, (eds.), Norwich, 1984.

10

The Frustrated Development of the Haitian Sugar Industry between 1915/18 and 1938/39: International Financial and Commercial Rivalries

Guy Pierre

> There are in Haiti large areas of land suitable for cane cultivation and the conditions for labour are particularly favourable. Nevertheless, the production of sugar for export in this country is made only by a company in Port–au–Prince.'[1]

The short period 1915–1918 marked an important moment in the economic development of Haiti, for the export of sugar, which had been the life blood of the economy before the revolutionary era of 1790–1820,[2] and had disappeared after this time, had a rebirth. This raised hopes of a great future for the country, but these hopes faded rapidly as the Haitian sugar boom ended with armistice in Europe. However, sugar continued to be produced for the domestic market, and there was also a revival of exports, interestingly enough, the strongest resurgence coming during the 1930s and reaching its peak in 1938/39. So, although the spectacular promise of the war–time sugar boom was not realized, for two decades the sugar industry was to enjoy a fair degree of prosperity.

Nonetheless, while sugar output and exports grew in the interwar years, on the whole Haiti did not experience the same dramatic expansion as other fellow Caribbean producers. Why was this so? Knight has argued that this was due to the fact that the peasants not only did not sell their land but also would not work under contract.[3] Other authors[4] have stressed that compared to Cuba or Central America the conditions for large capitalist plantations were not favourable in Haiti and this made the country

121

relatively unattractive to foreign capital. Another factor mitigating against the more extensive development of sugar, or so it is argued,[5] was the migration of Haitian workers to neighbouring countries, which in the period 1920–1930 led to a scarcity of labour in the country.

These are important and interesting theses, but as will be argued here, are essentially incorrect. It will be shown that despite peasant resistance and outflows of labour, there was in Haiti a shortage of neither land or labour. Instead, the industry's growth was held back because of rivalries between large international sugar corporations. The most central problem facing the industry was one of finding markets, coupled with the constraints imposed by the actions of the National City Bank of New York and the Haytian American Corporation (NCB–HAC), which as a financial–industrial group strictly controlled the Haitian sugar industry.

In this study,[6] we will look first at the reasons why US and British sugar interests could not move into Haitian sugar. The second issue to be considered is the discrete help given by the US Financial–Adviser – Receiver General to the NCB–HAC so that they could totally control the domestic market. Finally, the battle between various sugar groups to conquer the US market will be investigated. But before looking at these questions it is necessary to offer a picture of the growth of the sugar industry in Haiti during these years. To do this it is convenient to divide the entire period into three main sub–periods, 1915–1920, 1920–1929 and 1929–1939.

The cycles of development of the Haitian sugar industry 1915–1939

1915–1920: The sudden rebirth of the sugar economy

Despite evidence of some steam driven mills and large estates[7], before WWI the country's sugar industry was backward and relatively unimportant. Even the domestic market for sugar tended to be fragmented into small local markets. Nonetheless, it does seem that there was a degree of resurgence from 1910 in that the small, old–fashioned and mainly family–run production units known as *guildives* began to produce both more alcohol (which was their main product) and more *rapadou* (crude non–centrifugal sugar) for local needs.

The importance of the domestic market or markets became completely overshadowed with the outbreak of war, as the exceptional price rises which ruled in the world market after 1914 completely transformed the Haitian industry. For the first time in almost a hundred years a large external market opened up. Exports grew from almost nothing (236 kilos) in 1915 to over 4,000 metric tons in 1917/18, and by 1919/20 sugar exports were worth about $900,000. Notwithstanding the fact that this was less in real terms (about $400,000, if 1913 = 100), the war was a boom period for the industry, export earnings in constant prices increasing by an annual average of 152 per cent. By 1920 the Haitian sugar industry had been reintegrated into the world market.

1920/21–1928/29: crisis, difficulties and the consolidation of the sugar industry

The spectacular war and post–war price rise was caused by speculation in the London and New York markets by the large US and US–Cuban sugar interests. After Allied controls had been lifted this speculation drove the world sugar price to unprecedented levels. The collapse which followed was equally spectacular. In Haiti, the price of sugar fell from 26 cents (US) per pound to only 8 cents from 1920 to 1921. Although the quantity exported

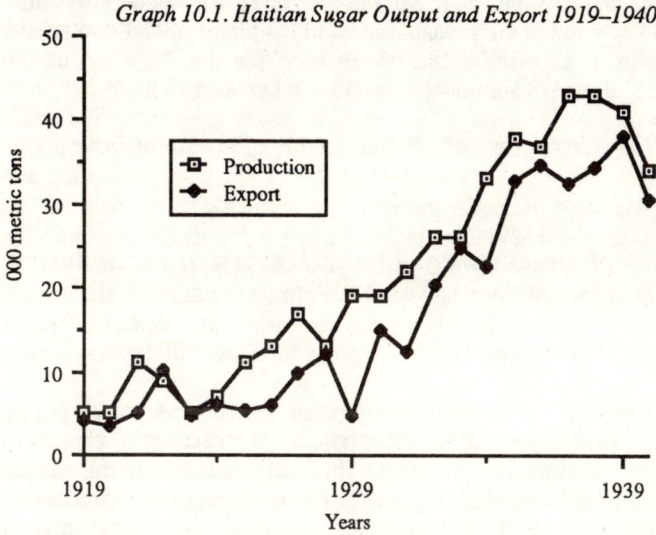

Graph 10.1. Haitian Sugar Output and Export 1919–1940.

Source: FAO, *The World Sugar Economy in Figures, 1880–1959*, (Rome, 1960)

increased by 50 per cent, exports earnings slumped to $460,000, about half the level of the previous year.

The next year (1922) was even more difficult, with the local price for sugar falling to 5 cents. To earn about the same from exports, 11,000 tonnes had to be exported as against only 5,000 in 1921. From this point to the campaign of 1928/29 the Haitian industry encountered numerous

difficulties. Partly this was due to the fall in exports in the years 1921–
1923 and their virtual stagnation from then until 1926.

These years also saw the base of the industry consolidated. There were
two important aspects to this, one international the other domestic. From
about 1924 exports began to increase slowly, than somewhat more quickly
reaching the high post–war levels by 1927/28. This growth had an
important impact on the wave of consolidation between 1924/5 and
1927/28. But at the local level there were forces making for consolidation
from at least 1922. The separate local markets were being replaced by a
more unified national market, which as will be explained, was
monopolistically controlled. At the same time the production of refined
and unrefined sugar for local consumption increased, from 539 tonnes in
1922 to 4,000 tonnes in 1929.[8] There was also a growth in sugar
derivatives, such as molasses and alcohol. The latter product became
particularly important, with new distilleries being established near Port–
au–Prince, the city of Saint–Marc and the Leogane plain. Alcohol was also
later to be a key issue in the peasant unrest in the plains of Cul de Sac and
Leogane, as well as offering the fiscal base for the Steniot Vincent
Government and the US administration in Haiti between 1930–39.

1929–1939: Crisis, expansion and decline of the sugar industry

The consolidation of the sugar industry was suddenly interrupted by the
collapse in prices in 1929. This led to a sharp contraction in exports to
4,000 tonnes, or the same level as in the year 1918/19, representing a 333
per cent drop in foreign sales since 1927/28. In financial terms the impact
of this crisis was greater than when the post–war boom ended in 1921, for
sugar exports now represented a larger proportion of the country's
earnings.

To meet this new crisis the Haitian sugar industry adopted a policy
employed by many larger scale sugar producers, it increased its exports in
an attempt to minimize its unit costs. This tactic was successful in that
the local price, following that in London, fell until about 1933, remained
stable from then until 1935/36 and then began to rise slowly up until
1939. Because of this the country earned, in both current and constant
terms, an increasing amount from her sugar exports. In fact, in terms of its
average rate of growth (13 per cent) and its relative contribution to the
total exports, sugar was more important in the 1930s than at any time
since 1915. There was also in the 1930s a significant increase in the
exports of other sugar products, such as molasses and non–centrifugal
sugars.

Domestic consumption also increased in these years. The only data we
have are for one sugar enterprise, but according to these between 1929 and
1939 the local market was increasing by about 4.8 per cent per year. It

must be said, however, that this was because of an increase in the consumption of non–refined sugar, for sales of refined sugar fell by 50 per cent in the same period.

With the outbreak of WWII the expansion of the sugar industry suddenly ended, and a new chapter in the industry's history began. This was due to the strong growth of domestic demand, the resultant local deficit of production,and the virtual isolation of Haiti from the international sugar market. Finally, for the country as a whole sugar began to be replaced by bananas as the major export sector.

The two great obstacles to the development of the sugar industry: international commercial and financial rivalries

We have seen that between the wars the Haitian sugar industry developed strongly and had a major impact on the entire economy. It helped greatly in the general process of capital accumulation and in the modernization of the relations of production. Nonetheless, qualitatively the industry in Haiti did not prosper all that much, especially when compared to some neighbouring countries. It regional terms, the development of the Haitian industry was very much an epiphenomenon, as even after its dramatic expansion the Haitian industry output represented only 1 per cent of the sugar produced in the region.

The reasons for the less than dramatic development of the Haitian sugar industry, as has been argued, can be put down essentially to the problems created for that industry by financial and commercial rivalries. These issues will be considered in turn, but before that it is necessary to discuss the structure of the sugar sector itself.

There were basically two groups of sugar producers in the country, as well as two large foreign owned enterprises with a character setting them apart from both these groups. The first was composed of various small manufactures, distilleries and *guildives*, spread throughout the country, but mainly near the capital. It is not clear exactly how many of these units there were but it seems that they had little capital and were small scale, artisanal enterprises producing non–centrifugal sugar and alcohol (*le clairin*) for nearby markets.

The second group was far more important. This was made up of a few relatively large manufactures and distillers, belonging to two or three prominent local families,[9] but which did function on the capitalist basis.[10] They used modern equipment and employed on average about 100 workers. Their output included refined sugar (although the degree of polarity is uncertain), non–centrifugal sugar, molasses and alcohol produced for regional markets. These enterprises also owned their own cane lands, but worked these in general under an archaic and somewhat varied non–wage labour system, which included some wage–labour component.

Almost all the aforementioned enterprises were near the capital. Their owners, for example, T.C. Brignac y M. Laroche,[11] were both politically and socially very powerful in this region. One of the owners, R.T. Auguste, was head of the army and became president of the republic in 1912. But the independent position of these producers was not to last. For reasons which remain unclear, in the 1920s the principal firms in the Cul du Sac region signed 10 year contracts tó sell their cane to the Haytian American Sugar Company (HASCO), a central mill which had been established in 1918. This was the major sugar producer in the entire country[12], making all the sugar exported from 1917 to 1939, as well as most of the refined and non–refined sugar sold in the local market.

It had been set up in December 1918, in Delaware with capital of $7,500,000 by Charles Steinheim, John A. Christie and Frank Corpoy. It was made part of the Haitian–American Corporation, a holding company, which also controlled three other important companies; Compagnie de Chemins de Fer de la Plaine du Cul de Sac (PCS), which they used to transport their sugar, La Compagnie haitienne du Wharf de Port–au–Prince, and La Compagnie d'Éclairage électrique des villes de Port–au–Prince et du Cap Haitien.

The central also had many of its own cane lands, on which they tried to establish a labour regime similar to that employed by local sugar growers, but this attempt had to be abandoned. In 1927 they directly controlled over 8,000 hectares of land, including the properties of Auguste and the Lespinasse family. This represented 10 per cent of the land in Haiti planted in cane.[13]

It must be emphasized that this enterprise was the sole buyer of cane. The state was a great help in allowing it to become an effective monopsony. This was done through a concessionary contract which gave the HASCO certain financial and customs advantages. For example, between 1935 and 1939, for the export of each 110 kilos of sugar only a small duty was charged. At the same time from 1928 a tax of 6 to 13.4 cents per litre was placed on alcohol.[14] This ruined the small producers of *clarin*, who were gradually forced to abandon production and sell their cane to the HASCO at prices determined by that company.[15] These advantages also made it difficult for the island's other central, the North Haytian Sugar Company, to extend its activities.

This last named enterprise was established by C.H. Wanser in the north of the country in 1920 with a capital of $185,000. It had about 2,000 hectares, but it seems only about 25 per cent of this was in cane. At the moment little is known about this company, its financial links or why it was unable to expand. Possibly it simply could not raise sufficient capital to increase its output, or perhaps it was blocked by the strong economic and political position of the HASCO. As we will see, the HASCO was ever vigilant to protect its position within Haiti.

During the entire period under review the HASCO fought to maintain its financial position in the country and opposed the setting up of any rival sugar interests. But the protection of its capital and profits was, as was common, a discrete not an open process.[16] It was done through diverse influences brought to bear mainly by the National City Bank and the Haitian–American Holding Company on the local US administration and the Division of Latin American Affairs of the State Department.

The clearest example of this type of activity was the successful campaign to stop the formation in 1919 of the Anglo–Haitian Sugar Company part of the West Indies Company, in effect a financial holding company. The plan was to set up a central in Haiti backed by capital of $10,000,000. In response to this, the US financial adviser in Haiti set out to investigate the proposed company and soon discovered that it had close links with the MacDonald financial group in the US, a group which had run foul of the Haitian authorities in 1912 for failing to pay $1,000,000 to the government.[17] It is not clear whether it was before or after the the US State Department became aware of these links, but the Anglo–Haitian project was soon abandoned. It is important to note that much of the information upon which these findings were based came from a US lawyer, R. Farnham, who had been involved in MacDonald's earlier activities in Haiti, but at this time was working for the HASCO.

Beside the problems of direct opposition, there was another factor which strongly mitigated against the setting up of any sugar company in Haiti. This was the inability of Haiti to conclude a special treaty with the United States. During WWI there were a number of attempts to see if this could be done, but the State Department was unresponsive. When the American International Corporation, who were to form the HASCO, were investigating the possibility of investing in Haiti their attorney, Farnham, pursued the question of a special relationship with the State Department. He got a rather ambiguous answer, which it seems he interpreted as a positive one, and partially on the strength of this misunderstanding the central was launched. This was to prove a very expensive mistake.

Other potential investors were more cautious. For example, at about the same time, the Chartered Company wanted to establish a central in the north of the country. It had gone so far as to receive promises from a group of landowners who offered about 12,000 hectares of irrigated land for cane cultivation, together with firm agreements to sell only to the new central. However, despite these favourable conditions, which were improved by the ability to draw on extremely cheap local labour, without preferential access to the US market the company felt that their investment would not be prudent and they withdrew from Haiti.

Although the HASCO was all powerful in Haiti, in the world market it and the country were virtually powerless. This was particularly evident in what might have been its most important market, the United States.

During the interwar period only a very small proportion of Haitian sugar found its way into this market. Unlike Cuba, there was never an economic accord signed while Haiti was under the direct political control of the US. The position became substantially worse in the 1930s when import quotas were imposed. Haiti was allowed to export only 300 tonnes, although this had been raised to over 450 tonnes by 1936. The country was, therefore, forced to look to other markets.

It turned to the only realistic alternative, the British Empire market, and to a less extent to nearby Curacao and the Virgin Islands. Britain became the major market throughout the 1930s, taking more than 75 per cent of total exports. The raw sugar was refined in the UK and than re-exported. But, the British market was not all that secure for preference was given to both the newly formed domestic beet industry and empire cane producers. Furthermore, there were more powerful sugar exporters, such as the Dominican Republic, who, also without privileged access to the US market, competed with Haiti. In terms of the world market Haiti seemed to be in a cul–du–sac, ironically enough the name of the area in Haiti in which the HASCO operated.

It is also apparent that the HASCO–National City Bank group did little to try to improve the market situation. Instead of fighting, its response was to produce well below capacity during the 1930s and even below the world market quota offered to Haiti under the 1937 International Sugar Agreement. But could the bank have helped find new or better markets? At this stage, we simply do not know. Perhaps it was thought that battling for the HASCO would have adversely affected the banks investments in Cuba and the Dominican Republic. But until further research is carried out this must remain only a hypothesis.

Conclusion

We have seen that between the two wars the Haitian sugar industry developed in a very complex manner. On the one hand the process of capital accumulation was intensified, with economic and social structures showing great change. Monetary and capitalist relations were also extended because of the growth of the industry, especially in the years 1928–1939.

However, sugar was reborn in 1915–1918, a period of enormous difficulties. If found itself facing an increasingly saturated world market from the early 1920's until late in the following decade. Because of this it could not prosper as its sister industries in Cuba and other parts of the Caribbean. But the failure of this industry should not be considered as the failure of North American imperialism as such in Haiti. It was more that the lack of dynamic expansion was the result of the failure of the American International Corporation. Despite the fact that this corporation was able to block the establishment of rivals in the country, it remained very much weaker than the other North American groups that had been set up in sugar in Cuba, Puerto Rico or the Dominican Republic. It could not get anywhere with the US Congress. Having a strong position here during the long Kondratieff cycle of 1890–1945, was decisive for US domestic producers and for North American corporations in the Caribbean. Haiti was simply frozen out. Because of this the capitalist transformation did not proceed apace here, while nearby countries were able to achieve a certain level of development.

The case of the Haitian sugar industry demonstrates that for all the questions of technical and structural change, much of the modern history of the sugar industry in the Caribbean can be seen to hinge on the the history of intrigues carried out by members of the US Congress who were either suborned by corporation lawyers or who had shares or other interests in these corporations. It is also the history of discrete machiavellian relations established by these corporations with highly placed men in the Latin American Division of the Department of State and with those other men who had the local posts of Financial Adviser–Receiver General.

Before ending it is necessary to modify somewhat the idea that the HASCO group failed in Haiti. It is said that during the years 1918–1939 the corporation suffered continual losses. This was the opinion of both the Financial Adviser–Receiver General and various authors. Furthermore, it is in line with the collapse of Haitian sugar prices in the export market over these years. However, the financial administration of the HASCO between the wars was much more complex. It is not enough to look only at the movement in external prices. What has to be considered is both foreign and domestic sales. The two experiences were completely different. As we have seen exports were problematical throughout, although there was some slight growth, but local sales were very 'juicy' for the HASCO. In effect they had a monopoly of the local market, and to aid this the Haitian government imposed a duty of 8 cents per kilo on refined sugar. So although the foreign market was not a success, the HASCO was able to charge high prices locally, and this must have helped it to balance its overall financial position. However, only a detailed study of the records of this corporation will allow us to confirm this hypothesis.

References

1. *Informe anual, 1928–29*, Puerto Prince, Haiti, 1929.

2. On this early period see Paul Moral, *Le paysan haitien*, Port–au–Prince, 1978.

3. Melvin M. Knight, *Los americanos en Santo Domingo*, Santo Domingo, 1939.

4. Gerard Pierre–Charles, *La economía haitiana y su vía de desarrollo*, Mexico, 1984. Marion Leopold, 'Resistencia campesina y lucha de clases en Haiti', *Mexico Agraria*, xiv, no.1, 151–178.

5. Suzy Castor, *La occupación norteamericana en Haiti y sus consequencias (1915–1934)*, Mexico, 1971. Also Susy Castro, 'El impacto de la occupación norteamericana en Haiti y Santo Domingo', *Política y sociología en Haiti y Santo Domingo*, Mexico, 1974.

6. Being unable to obtain access to the archives of either the National City Bank or the Haytian American Corp., this study is based largely on the archives of the US Department of State [ADE.830.50].

7. Guy Pierre, 'La crise de 1929 et le developpement du capitalisme en Haiti', unpublished paper.

8. Albert Viton, *Aspects économique du développement de la production-sucrière en Haiti*, Rome, 1952.

9. Candelon Rigaud, *Promenades dans les campagnes d'Haiti*, Paris, 1928.

10. *Ibid.*. Turnier, *La société*.

11. Rigaud, *Promenades*.

12. ADE. 838.61351 H 33/10.

13. Rigaud, *Promenades*.

14. Millet, K., *Les paysans*.

15. Moral, *Le paysan haitien*.

16. We have not found any document in the archives of the US State Department the shows either implicitly or explicitly that HASCO directly intervened at this time.

17. Pierre, 'La crise'. Also see the contract signed by MacDonald in US Archives, Roll 85, File 838.6156 M14.

11

Technology and the Plantation Labour Supply: The Case of Queensland, Hawaii, Louisiana and Cuba

Edward D. Beechert

Introduction

Sugarcane cultivation is often contrasted with the more 'progressive' and more technologically advanced milling and refining of sugar. It is frequently assumed that cultivation and harvesting are low skill tasks. Technical change is thus minimized on the grounds that an unchanging and cheap labour supply made unnecessary any technological improvement in cane agriculture. The use of 'cheap' labour is assumed to obviate the need for labour saving practices and techniques.

This labour aspect of sugar production has been described variously as a function of imperialism, unequal development, or simply characterized as a retarding influence because of the plentiful labour. Plantation economies are seen as disadvantaged with respect to 'both the incentives and capability of adopting new technology.'[1] Technological advances are seen as reallocations of resources in order to avoid losses from labour disruption.[2]

It is argued here that there was a continuous effort to transform sugar cane agriculture beginning in the 1880s, which only reached an effective technical level of application in the post–1946 era. The inter–war period saw the maturing of a variety of technologies which presented new opportunities for mechanical solutions to agricultural problems. The developments in varietal strains, fertilization, irrigation, and cultivation were the most significant. Cane harvesting and transport changes were more difficult and slower to develop. The transformation of these

131

agricultural factors was shaped as much by non–technical factors as by the availability of technology.

Impeding circumstances will be examined in four major sugar producing areas: Queensland, Hawaii, Louisiana, and Cuba. In each of the areas, labour supply was increasingly tenuous and subject to rising wage levels. However, each area faced very different labour supply problems and the technological response varied widely in each case. What factors contributed to technological developments in cane agriculture? How were these developments related to questions of labour supply and labour costs? What environmental and ecological factors dictated technical usage? What improvements in agricultural technique were put into place and what impact did these changes have on labour costs and profitability?

Elements of plantation production

The analysis of plantation production has been dominated by the notion that the system is not possible without non–indigenous, tightly controlled, low wage labour.[3] It is even asserted that 'no plantation economy ever achieved modern economic development...[T]he failure of economic development in plantation economies flows from the fact that such a society is relatively disadvantaged with respect both to the incentive and capability of adopting new technology.'[4]

Sugar, unlike other plantation crops, presents a greater variety of environmental and economic obstacles to efficient and profitable production. The struggle to 'reconcile labour–effective cultivation units with efficient mill operation...has developed a wider variety of organizational forms than in the production of any other plantation crop.'[5]

When sugar producing areas are compared on the points of labour supply and deployment and technological development it becomes apparent that factors other than the basic plantation system are responsible for the observable differences. The different plantation systems derive from ecological factors and political economies rather than from factors of labour supply and technology.

The difficulty in establishing a general definition of plantation that will meet the range of conditions is clear. Local conditions, regardless of the homogeneous nature of plantations, will 'bend the productive goals of the enterprise. The social relations of production will emerge from much more than the fact of large scale enterprise.'[6] The specific character of modification cannot be predicted and it is difficult to generalize from one example to another in terms of world 'market mechanisms'.

A non–indigenous labour supply, common to many plantation areas in the 19th and 20th centuries, did not always mean a low cost labour force. Political circumstances outside the plantation often altered supply characteristics. Worker resistance frequently altered the circumstances of supply. Even when indenture systems provided a large measure of 'control' at the point of production, the conditions surrounding the continued importation and the continued utilization of the labour force altered the social relations of production and, eventually, the mode of production.[7]

Milling and refining were the first sectors of sugar production to be developed. Drawing stimulus from beet sugar processing, cane milling and refining quickly reached a high level of efficient extraction and refining.[8] The spread of scientific research was rapid and worldwide. A variety of sugar journals featured information on all aspects of sugar production throughout the sugar producing world. A consistent theme in these journals was the imperative need for technical improvements, both in milling and in agriculture. News about new varieties, implements, and above all, news on harvesting machines were regular features.

Cultural implements appeared on the market, many were quickly adapted to sugar cultivation where circumstances permitted. Thus, after their introduction in 1875, use of the riding cultivator and disc plough spread rapidly in Louisiana. The disc plough and harrow and the double disc riding cultivator increased cane yields in Louisiana to over 'ten tons of cane per acre, and the sugar product, 700 pounds.'[9]

Growing competition for sugar markets stimulated the search for efficiency. The director of the Louisiana Sugar Experiment Station warned, 'In this day of strong competition...every possible economy must be practiced. Only by the adoption of the most improved and economical methods can the tropical cane be maintained.'[10] As with other farm implements, cane cultivating machinery tended to be developed out of practical experience before being developed commercially. Introduction of new machinery, however, raises a serious problem. Change in one component of a system creates a stimulus or need for changes elsewhere in the system. 'Awareness of imbalances between components has continually led to an exploration of possibilities for corrective action whose eventual result was a major improvement in productivity.'[11]

The limitation of technological improvement in cane agriculture is basically in the sequential nature of cane–sugar production. The vulnerability of cut cane places limits on its transport. Mills have limitations of capacity. These factors must be kept in a rather narrow balance lest imbalances in production cause costly production losses.

Overcoming the problems of cost, durability, and practicability can be a very long process. The search for a cane harvester took about 100 years from the first U.S. patent in 1854 until the appearance of a practical, workable harvester in 1952. It was thirty years before a workable, durable corn harvester, first patented in 1850, was available.[12] Early patents of corn harvesters were claimed as sugar cane harvesters, apparently seeing an identity between Louisiana sugar cane and corn stalks (e.g. US patent #24,992, 1854). The first explicit cane machine was patented in 1860, (patent #30,266). Some 350 patents later, a workable cane harvester cut a portion of the 1931 Florida crop. It would be another twenty years before machines capable of dealing with the general conditions of cane production would be available and could substitute for the skilled hand cutter.[13]

The sequential nature of technological development is demonstrated in the technology of agricultural equipment. The metallurgy which could produce steels capable of meeting the needs of cane agriculture evolved only in the late nineteenth century. Bessemer steel, although a great advance in steel making, was too brittle to withstand the many stresses of cultivation and heavy soils. The post–Bessemer steels, such as open hearth steel, were required to create the flexibility and durability needed for cane tools.[14]

The search for agricultural innovation

Queensland cut off the supply of indentured labour as a deliberate racial policy. A system evolved of small cane farming operations employing white harvesting gangs, which was supported by a government controlled central milling system. The harvest crews had to be supplemented by European immigration, even in the face of vigorous opposition to the use of foreign–born white workers.[15] The necessity of relying upon a limited labour force put pressure on the cane farmer of Queensland to seek mechanical solutions. The cost and capacity of the harvesting machines made them unprofitable for small farmers. Cane had to be harvested according to a rigid schedule to avoid disrupting milling schedules. Mechanical problems were secondary to mill and transport capacities. Expensive mechanical solutions to the labour shortage were out of the question. The Queensland small cane farmer had to seek out solutions in the realm of home–built machines such as harrows and cane planting sleds. Varietal canes, irrigation and fertilizing techniques offered greater opportunities for profitable production than complicated mechanical devices.

In the case of Cuba, the production unit was less well– defined than that of Queensland. The development of the '*colono*' system of cane production and central milling introduced a variety of participants. Some landowners produced cane as a *colono* for central mills, others rented land and produced cane, others worked on a share–cropping basis.[16] The appearance of the large American owned mills after 1900 exacerbated this older division of production.

The dizzy expansion of Cuban production between 1900 and 1921 created a situation in which there was a great variation in the quality of production, ranging from cane planted among newly felled trees to the machine–cultivation of land under control of the large mills, the so–called administration cane land.[17] The dominant mode of production in this period was on crudely cultivated land, using oxen power, with little or no fertilization, no irrigation, and with low yields. This was in contrast to the highly technical production on the large estates.[18] Total costs for oxen cultivation came to $370 per *caballeria*, as compared to $161 for tractor cultivation. The saving on 60 *caballerias* would pay for two 40 hp tractors, two disc ploughs and a 16 disc harrow.[19]

After the crash of 1921, the question for all Cuban producers became not one of efficiency or technology but one of survival – whose cane would be harvested under the quota system? The political clout of the *colono* turned out to be more effective than that of the capital intensive American operations.[20] While American owned mills put a great effort into modernization, ordering the latest tractors such as the Storey Gyrotiller, small motorized cultivators and the like, access to milling quotas was determined on a political basis.

The short growing and harvesting season in Louisiana created special problems in the political economy of sugar production. With cane generally planted in February and harvested in October and November, the planters had different pressures than those felt in Cuba, Queensland and Hawaii with their longer term cane. The nine to ten month season meant that Louisiana was always harvesting immature cane and facing the possibility of loss through early frosts each year.

The short seasons created special problems for the Louisiana grower who was vulnerable to severe loss in the event of a labour shortage. The burdensome nature of the work and the lure of higher paying jobs elsewhere combined to create a continuing problem of labour recruitment and a continual search for mechanical solutions. As early as 1886, the planters offered a prize of $1,000 for a mechanical harvester. The motivation was the 'scarcity and cost of field labour in Louisiana.' The perennial problem generated pressure to 'grasp any mechanical invention' which would ease the labour shortage.[21]

Almost as an annual ritual growers discussed importing workers from Mexico or the passage of legislation permitting the 'rounding up of idle Negroes and inform [them] they have to go to work on the plantation or go to jail.'[22] After spending some $50,000 for labour recruitment in 1904, the Louisiana Planters' Association urged the development of a 'combined device for both cutting and loading' as an urgent necessity. They recommended spending half the annual sum for labour recruiting for developing a combined machine. Between 1917 and 1930, of fourteen patents granted to Louisiana developers alone, only two progressed beyond the experimental stage. Equally limiting was the financial state of the sugar industry. Falling sugar cane production, partly due to crop disease, lessened the enthusiasm of the planters for heavy investment in machinery.

The industry did develop machines of considerable significance. The Castagnos cane loader was developed by 1915 eliminating one of the more burdensome tasks. The machine was capable of being constructed by local shops and was low in cost.[23]

The recovery of Southern sugar through new cane varieties arrived with the Falkiner Cane Harvester and the Depression. By 1927, the Falkiner harvester had reached a state of development that warranted commercial production. Entering into arrangements with Allis Chalmers, the machine was put into production and destined for use on American lands in Cuba. Because of the political turmoil there, the machines were diverted to the newly formed Southern Sugar Company at Clewiston, Florida. In 1931 a significant portion of the crop was cut by machine. The Depression and the heavy unemployment made it inexpedient to continue the use of the machines and they were set aside. Efficiency gave way to political reality.

From the point of inclusion of Hawaii in the American market in 1876, sugar producers of Hawaii had substantially different conditions than those of other areas. For one thing, the number of producers was dramatically and steadily reduced until the industry was represented by five agencies, each controlling a string of sugar companies. To a degree unknown elsewhere, the Hawaiian Sugar Planters' Association came to be the policy maker and coordinator of research and marketing activities.

A labour supply was ever a critical problem for Hawaiian producers. Indigenous workers were insufficient to support the scale of industry which evolved. Imported labour was therefore imperative. These supplies were always subject to the vagaries of politics, both domestic and international. Frequent shortages of workers fueled the search for mechanical aids. In 1904, the Association set up a Labor Saving Devices Committee which was charged with the task of reducing hand labour. 'What is needed most is apparatus and machines which will cut out large percentages of our hand labour in our fields...'.[24] Even before the Committee was formed, work had begun in 1902 on the development of a cane loading machine.

The wide variety of soils and terrain in Hawaii created a myriad of problems of varying technical complexity. The one consistent element in Hawaiian cane production was the heavy weight of the cane. With a growing season of from eighteen to twenty–four months, in either irrigated or the unirrigated plantations, the problems of harvesting and loading were severe. Some plantations in areas of the heavy rainfall could take advantage of the slope and ample water supply to transport cane by water flumes directly to the mills, or in some cases to mainline railroads. Seven plantations on the island of Hawaii used wire cable transport to send their cane to the railroad.

Hawaii carefully monitored the work being done on cane harvesters in Queensland, Cuba and Louisiana. However, they found none that could cope with the heavy, tangled cane of their plantations. In many cases, machines would work only on level fields, free of rocks. Most of Hawaii's sugar fields failed to meet this test. Skilled hand cutting remained for many years the most practical and efficient means of harvesting heavy cane.

Although heavy machinery, loaders, tractors, and harvesters easily dominated the attention given to technology, perhaps more significant in the long run was the attention given by the HSPA Experiment Station to matters such as irrigation and fertilizers and constant varietal and pest control research. As early as 1914, irrigation experiments demonstrated savings in labour inputs. Water courses on each side of the row required one irrigator per 8.29 acres. By irrigating alternate furrows, the figure was increased by 1916 to 12 acres per man.[25] The straight line system labour requirement for some plantations was 19 man–days as compared to the 50–60 man–days required under the older contour system.[26] The HSPA estimated that changes in harvesting, such as abandonment of set times for ratooning and the adoption of 'cut and grow' practices increased the efficiency of labour and machinery. 'Better coordination of operations such as fertilizing and irrigating accounts for about 20–25 per cent of the increased yield. Better fertilization practice contributed about 15–20 per cent of the increase. As much as 10–15 per cent of increased yield was due to effective pest control.'[27]

Ironically, some of the increased yields obtained from the improvements in cultivation, chemical weed control, soil improvement practices, and new varieties which had been achieved by 1930 were partially lost in subsequent years. In 1934 Hawaii suffered a double blow. The Philippines Independence Act cut off the supply of Filipino labour and rigid quotas on sugar production were imposed.

A hasty solution was the introduction of grab–rake harvesting, with resulting losses estimated to be as high as twenty per cent of the cane grown.[28] Grab harvesting, introduced in 1937 in response to an absolute shortage of labour, produced heavy loads of dirt and trash which had to be removed at the mill before crushing. One extreme example of the dirt and trash problem was Lihue Plantation where 397,082 tons of harvested cane brought to the mill contained 123,062 tons of dirt, trash, and damaged cane.[29] As an expression of the desperate search for mechanical aids, the example of the Le Tourneau Harvester in 1939 would be difficult to exceed. The machine put into the field weighed 39 tons, with tyres 10 feet in diameter. The large machine quickly demonstrated to HSPA engineers that extreme size and brute power would not efficiently harvest cane. Although inefficient, the grab rake was at least workable.[30] Despite the drawbacks of mechanical harvesting, the Hawaiian work force was drastically reduced. In 1939, some 41 per cent of the work force was engaged in harvesting. By 1946, only 17 per cent were employed in the same task.[31]

While the industry was increasing the cane yield from an average of 69.1 tons per acre in 1932 to 71.3 tons of cane per acre in 1946, yield of sugar per ton was falling. In 1932 the yield was 248 pounds of sugar per ton of cane. In 1946 this figure had fallen to 227 pounds of sugar. Twenty per cent more cane was, in effect, producing nine per cent less sugar. The fall in yield resulted basically from the damage inflicted in harvesting and transporting.

Conclusions

It can be argued that three basic conditions have transformed the world sugar industry since 1918. These are: (a) the introduction of new varieties in the 1920s; (b) the evolution of trucks and tractors to a practical, economical level where they could reliably replace animal power; (c) A critical shortage of labour in Queensland, Hawaii, and Louisiana.

Technology was freely available to those who chose to apply it in each of the major areas of sugar production. Capital, except in Queensland and Hawaii, was not always available to those needing it. The protected nature of Australian sugar ensured a level of support for the cane producer and refiner. Hawaii's plantations, enjoying tight, inter–locking corporate control and a generous subsidy were well able to commit the capital required for technological development. Large scale operations, under central direction, with full control over all facets of production from cultivation to marketing, as well as financial control created something approaching ideal conditions. With the sole limitation of production quotas, Hawaii continued during the period to enjoy optimum conditions.

The worldwide restriction of sugar markets cut Cuba's production to less than half its earlier production. Cuba, facing increasing domestic turmoil, rising nationalism, and a loss of world market, resolved its problems in part by reducing the share of the more efficient, highly capitalized American mills. Given the threat of the Platt Amendment and the record of United States intervention, one would not have predicted that the inefficient *colono*–producers would be able to win a larger share of the available quota for cane. Political realties, not technological considerations prevailed here.

The widely varying physical and social–economic conditions of the four sugar producing areas made it impossible to simply transfer technology from one area to another. Beyond questions of labour supply were more important questions relating to the political economy of each area.

One result of the world–wide mechanization of sugar production has been to put pressure on the small producers. In Queensland, the higher capital requirements now prevailing appear to increase the tendency toward consolidation. Louisiana producers have always had the option to divert their land to other crops on a yearly basis, an option not open to the other producers. Subsidized crops such as soy beans and cotton offer attractive alternatives to sugar production. In Hawaii, capital has begun to move out of the sugar operations as the controlling corporations shift their focus to other activities, such as land development and tourism. Unless sugar prices and/or subsidies increase, Hawaiian plantations will be further mechanized and reduced in number.

The mechanization which began as a partial solution to problems of available labour supply has now moved beyond that stage to a transformation of production in each of the areas. Cuba has moved to the revolutionary solution for its problems, a move which dictates a rapid acceleration of mechanization and increase in efficiency. Queensland, like Hawaii, is gradually being pushed back into a system of large units of production to achieve economies of scale. In all of the areas, including the many not considered, the relevant factors turn out to be something other than the traditional notions of plantation production. The solutions required were not entirely matters of labour supply or technology but dependent on political and social realities.

References

1. Jay Mandle, 'Patterns of Caribbean Development', *Caribbean Studies*, 1982, 2, 46.

2. Louis Ferlerger, 'Farm Mechanization in the Southern Sugar Sector After the Civil War', *Louisiana History*, 1982, 23, 33–4; See also, Ferlerger, 'Capital Goods and Southern Economic Development', *Journal of Economic History*, xlv, 2 June 1985, 417; James Geschwender, 'The Hawaiian Transformation: Class, Submerged Nation, and National Minorities', in Edward Friedman (ed.), *Ascent and Decline in the World-System* , Beverly Hills, 1982, 5, 189–90; James Geschwender and Rhonda Levine, 'The Rationalization of Sugar Production in Hawaii, 1946–1960: A Dimension of the Class Struggle, the Hawaiian Sugar Industry', *Social Problems*, Feb. 1981, 30, 358; George Beckford, *Persistent Poverty: Underdevelopment In Plantation Economies Of The Third World* , New York, 1972, 60ff.

3. Jay Mandle, *The Roots Of Black Poverty,* , Durham, 1978, 10; Beckford, *Persistent Poverty*, 60ff.

4. Mandle, 'Patterns', 46.

5. Philip Courtenay, *Plantation Agriculture*, Boulder, 1980, 128–29.

6. Sidney Mintz, 'The Plantation As A Socio–Cultural Type', in S. Mintz, E. Wolf (eds.), *Seminar On New World Plantations*, Mona, Jamaica, 1957, 49.

7. Edward Beechert, *Working In Hawaii: A Labor History*; Honolulu, 1985; Kusha Haraksingh, 'Labour, Technology and the Sugar Estates in Trinidad, 1870–1914', in Bill Albert and Adrian Graves, (eds.), *Crisis And Change In The International Sugar Economy, 1886–1914*, Norwich and Edinburgh, 1984; Brij Lall, 'Girmitiyas: The Origins Of The Fiji Indians', *Journal of the Pacific History*, 1984.

8. A. C. Barnes, *The Sugar Cane*, London, 1964, 416–20.

9. W. C. Stubbs, *Sugar Cane: A Treatise On the History, Botany and Agriculture of Sugar Cane*, London, 1897, 145–46.

10. *Ibid.*, 154.

11. Nathan Rosenberg, 'The Direction of Technological and Change: Inducement Mechanisms and Focussing Devices', *Economic Development And Cultural Change*, October 1969, 18, 9.

12. Robert Ardrey, *American Agricultural Implements*, New York, 1894, 115.

13. A C. Barnes, *Agriculture Of The Sugar Cane*, London, 258–59; 'A Wool Grower in the Cane Field', *Australian Canegrower*, Dec. 1983, Jan.& Feb. 1984.

14. Nathan Rosenberg, *Inside The Black Box Technology and Economics*, New York, 1982, 91–92, 146.

15. Adrian Graves, 'Crisis and Change in the Queensland Sugar Industry, 1862–1906', in Albert and Graves, *Crisis and Change*, 278–79; Harry T. Easterby, *The Queensland Sugar Industry*, Brisbane, 1933, 28.

16. Brian Pollitt, 'The Cuban Sugar Industry and the Great Depression', *Bulletin Of Latin American Research*, 23:4, 1984, 9–10.

17. *International Sugar Journal [ISJ]*, 1923, 147–48, 512; Pollitt, 'Cuban Sugar Industry', 12–13.

18. C. J. Robertson, *World Sugar Production And Consumption*, London, 1934, 28–9.

19. *ISJ*, Jan. 1937, 16.

20. Pollitt, 'Cuban Sugar Industry', 3.

21. *ISJ*, Jan. 1907, 44–5.

22. *Facts About Sugar [Sugar y Azucar]*, *(FAS)*, 31 Oct. 1925, 1041.

23. *ISJ*, 1915, 24; FAS, 8 Jan. 1927, 39; Hawaiian Sugar Planters' Association (HSPA), *Proceedings Of The Annual Meeting* (LSD: Labor Saving Devices Committee; CLT: Cultivating, Loading And Transport Committee (1940), 28.

24. HSPA, *Proceedings*, 1904, 244–45.

25. HSPA, *Proceedings*, 1916, 56.

26. *FAS*, August 1932, 321.

27. HSPA, *Proceedings*, 1952, 39–40.

28. *Ibid.*, 39–41.

29. *Queensland Cane Growers Quarterly Bulletin*, 1 Oct. 1946, 77–9.

30. HSPA, *Proceedings*, 1939, 14.

31. James Shoemaker, *The Hawaiian Economy*, Washington, D.C., 1947, 57–8.

12

Crisis and Change in the Australian Sugar Industry, 1914–1939

Adrian Graves

Introduction

Despite the turmoil of the international sugar economy and the onset of the Great Depression, the interwar period saw the Queensland sugar industry achieve a remarkable rate of growth accompanied by very significant advances in capacity and productive efficiency. Raw sugar output expanded by nearly 270 per cent over the period and exports rocketed from a few thousand tonnes to around 400,000 tonnes, accounting for about 5.7 per cent of world trade in raw cane sugar at the outbreak of WWII.[1] The key to this success was the extraordinarily intense involvement of government in the industry both at the provincial (or State) and the Federal levels.[2] This essay attempts to survey, explain and evaluate the role of the state in the Queensland sugar industry in the interwar years. But first, it is necessary to establish briefly the historical background to the period under examination.

State and industry before WWI

Sugar was first manufactured in Queensland in 1862.[3] Until about 1890, the industry was based almost exclusively upon an extensive system of plantations which relied upon a field labour force of immigrant indentured workers recruited from the islands of the south–west Pacific. State involvement in the industry was limited to regulating the labour trade until the 1880s when the industry was plunged into a deep crisis of capital accumulation. The subsequent reconstruction of the industry was orchestrated largely by the colonial state through laws which encouraged cooperative central milling, the establishment of sugar experiment stations and liberal land legislation. The newly formed Commonwealth Parliament became directly involved in the industry at the outset. It abolished the labour trade in 1901, which had by then become redundant to the requirements of the reconstructed industry and established tariff and excise duties which both protected the home industry and generously subsidised it. The Commonwealth Government's involvement had not only established the principle of federal intervention in the industry but also the use of state policy to engineer changes in the sugar industry, its structure and its pattern of regional development. The effects of the intervention were marked, as the industry experienced a rapid rate of growth in the years immediately following Federation.[4] At the same time, the Queensland Government continued to increase its budget on the scientific aspects of the industry, promote further central mills and enact relevant (inadequate) labour legislation.

The bifurcated intervention of State and Federal Authorities in the industry's affairs was not without its problems. In particular the excise and bounty policy (which did not benefit millers and refiners directly) and the Federal authority's increasing intervention in wages' issues (which raised production costs) were a source of some tension between the State and Federal Governments. The complicated contradictions and conflicts within the industry itself, however, between workers, growers, millers, refiners and distributors and significant consumer dissatisfaction with high sugar prices were also expressed in the political sphere. These tensions and an extremely bitter sugar workers' strike in 1911, gave rise to a searching inquiry into the industry by the Commonwealth Government in 1912 which attempted to sort out some of these difficulties.

The upshot of the Commission was an even greater level of state intervention in the industry, through its continued protection and subsidization; the establishment of measures to placate the internal conflicts within sugar production through rationalizing the system of cane purchasing and other aspects of the relationships between farmers, millers and refiners; improving sugar workers' pay and conditions; and the provision of privileged access to sugar output by allied industries. To placate the now intense consumer hostility to the protection of the industry, the Commission's recommendations were justified in terms not only of the crucial importance of the industry to the regional and national economies but on the grounds of Australia's national security, as a cost of maintaining the White Australia Policy, and 'from the point of view of defence (to effect) the settlement and cultivation of the tropical and semi-tropical areas of the...continent'[5] The appeal to racist and nationalist arguments was an extremely potent political device in the Australian context and the industry employed them persistently over the following decades to justify the high level of state protection it enjoyed. For the outbreak of WWI dramatically broadened the forms of state intervention in the Australian sugar industry.

WWI

The onset of WWI prompted the state to assume full control of the Australian industry in so far as the purchase of raw sugar, its transport to refineries, its manufacture into refined products and the sale of refined products were concerned. This control was effected through proclamations under the Customs Act by which embargoes were imposed upon both imports and exports of sugar and by the enactment of the Sugar Acquisition Act of 1915 whereby the Queensland Government acquired the total Queensland sugar output and then sold it to the Commonwealth Government at an agreed price. The much smaller amount of cane sugar produced in New South Wales was sold to the Federal Government by the Colonial Sugar Refining Company Limited (CSR) at the same price. Rather than set up a separate authority to handle the sugar it purchased from the Queensland Government, the Commonwealth appointed CSR and the other, albeit insignificant, Australian refiner, the Millaquin Sugar Company Limited, to act as its agents. These companies arranged for the transport of the raw sugar from the mills to the various refineries, for its refining and then for its sale to the consumers, as refined sugar, treacle and golden syrup. As to prices, the Federal Government itself not only set the price of the raw sugar it purchased from the industry, it also fixed the wholesale, manufacturing and retail price of refined sugars distributed on the domestic market.

The official justification for these drastic interventionist measures was threefold; (a) to protect Australian consumers from possible price exploitation by the local producers due to the rising international price of sugar occasioned by the war; (b) to ensure adequate supplies of sugar to all Australian consumers during the war by preventing the Australian sugar producers from exporting their sugar to capitalize on high prices abroad; and (c) to protect those engaged in the industry from the operations of State based Food Prices Boards.[6]

The first two justifications did not please the industry in the first instance, since its natural instinct was to take advantage of the favourable international prices. It was apparently no consolation to the producers that the state had guaranteed them exclusive access to the domestic market. On the other hand, nullifying the impact of the Food Prices Boards of Victoria and New South Wales was an advantage since the FPBs had set the price of sugar well below current market rates. The industry was adamant that it saw the problem of sugar pricing as critical to its survival. But the major problem facing the Government was how to calculate a price to satisfy all the competing interests. In the end, a number of factors were taken into account in fixing the purchase price of raw sugar after 1915. The most consistent elements in the calculation, however, were the costs of production and cost of living rises. Under the rubric of production costs, the Federal Government paid most attention to wages' costs.

Steep wage rises awarded by both State and Federal wages' tribunals following the settlement of the 1911 sugar workers' strike had contributed substantially to increases in the cost of production for which, the industry argued, it should be properly compensated. It is also important to understand that these awards did not derive merely from industry conditions but were to some extent attributable to the Australian policy of industrial arbitration and conciliation under the Commonwealth Arbitration and Conciliation Act of 1904, and the famous 'Harvester' judgement of Mr Justice Higgins of 1907, which first established the Australian standard wage based on the cost of living. The impact of the wage rises on the Federal Government's price fixing policies during the war period on both the retail and wholesale price of sugars can be observed by reference to Graph 12.1.

It was also within this period that the first serious cane pricing policy and administrative structure were set up. These, however, were developed under the control of the Queensland Government through a series of Acts passed between 1915 and 1922.[7] These laws attempted to regulate the contentious issue of cane disposal by fixing the price and conditions governing the delivery of cane to mills through Sugar Cane Boards which met annually for the purpose. As a first step, the Boards assigned all lands owned by cane growers who were supplying cane to particular mills in each of the years 1915 and 1916. Local Cane Prices boards were set up to service the assigned mills. Assigned land included all the land of the assignee, whether it was in cane or not. Of great significance in the twenties and thirties, as we will observe later, the assignment of land was considered annually.

As to cane prices, the Boards fixed this according to its sweetness calculated on the basis of the canes' commercial cane sugar content (ccs). Another factor taken into account in setting cane prices however, was a minimum standard of mill efficiency, to assure the grower that a reasonable amount of sugar was recovered in the manufacturing process. Apart from leading to an increase in the value of cane which was delivered to the mills, growers enjoyed other benefits under the Acts, including better conditions of delivery, payment by the mills for railway and mill haulage, the compulsory arbitration of disputes between grower and miller, and a voice in the declaration of deductions on inferior cane.[8]

The ultimate purpose of these acts was to obviate grower and miller conflict, an objective they largely achieved. Associated with this, however, was the further aim of institutionalizing incentives to encourage cultivation and milling efficiency. Thus growers were rewarded for the sugar content of their cane. Equally, millers were encouraged to optimize mill efficiency, not only through the operation of the pricing system itself, but through a structure of cane testers and inspectors appointed ostensibly to protect the growers, whose duties included the supervision of weighing, sampling and the chemical analysis of cane.

Although intra–industry conflict was lessened somewhat by these policies, they did not lead immediately to significant increases in output nor to improved efficiency in sugar production. This was attributable to some extent to particularly adverse seasons during the war. But there is also evidence to suggest that producers did not invest in production during the war, and in its aftermath. This was explained later as being due in part to the financial institutions' reluctance to advance loans to millers while the state had such extensive powers over their affairs, and in particular, its 'restrictive legislation'.[9] By contrast, cane farmers, soon after the war at least, enjoyed good access to bank credit[10], but here, lack of investment was put down to a pessimistic outlook due to the adverse price of sugar in Australia compared with those obtaining on world markets.[11]

Australian consumers gained from relatively secure supplies of sugar during the war period, but they paid high prices, even if they were still lower than much of the rest of the world. The real winners, however, were the allied industries, who enjoyed access to very low, concessionary prices of sugar by world standards, secure supplies, and a total absence of competition from abroad. Not only did the war disrupt international trade in jams, canned fruits, biscuits and confectionery, but also the importation into Australia of confectionery, chocolates and biscuits was banned from 10th August, 1917. About the same time, the Federal Government set a very attractive concessionary price for refined sugar for use in bond to manufacture jams, jellies and canned fruits for export. This measure was designed specifically to encourage the Australian fruit industry. These concessions contributed to a remarkable expansion of output in all the allied products, and to a sharp increase in the export trade of jams, canned fruits and condensed milk. In particular, huge profits accrued to these manufacturers during the war and in the immediate post–war period on their exports of large quantities of foodstuffs containing sugar to the British, Australian and allied armies.[12]

The 1920s

State controls over the sugar industry became a major issue after the war. The immediate problem was to what extent should the state continue its involvement with the industry under the peace time conditions. A Royal Commission appointed in 1919 to investigate the industry, proposed rationalizing state involvement under the authority of a Commonwealth Sugar Control Authority which was to take complete responsibility for industrial arbitration, efficiency in cultivation and milling, cane and raw prices, conditions under which cane was delivered to the mills, efficiency of sugar mills, raw sugar prices, recommendations for new areas suitable for cane cultivation, mill sites, and control of sugar experiment bureaux and stations.[13]

The problem in effecting this proposal was that a recent Federal referendum which proposed greater Commonwealth powers over domestic commerce and industry had failed, although it received overwhelming support in Queensland. At the same time, there was a great deal of consumer pressure, especially in Melbourne, where the Federal Parliament was then sited, for the state to withdraw entirely from its involvement with the sugar industry. Consumers argued that the industry enjoyed unprecedented protection, was a burden on the taxpayers, did not encourage efficiency and resulted in unreasonably high profit rates for the producers, in particular CSR. Profiteering was always the rallying cry of the consumer organizations throughout the interwar period, and in putting its case, the industry devoted great attention to defusing that charge.

While these arguments were taking place, however, the state was forced into considering the price of sugar over which it still retained its wartime controls. By early 1920, the industry had convinced the Federal authorities that it was in serious trouble. The low fixed price of sugar was not encouraging producers to invest, particularly on the milling side. At the same time, demand was rising, and the state was forced to import greater quantities of expensive sugar, which it sold on the domestic market at home prices. The substantial subsidy the state was thereby forced to impose on the imported sugars predisposed it to the industry's call for a substantial increase in the domestic price of sugar. The impact of this on the price of raw and wholesale sugars can be observed by reference to Graph 12.1.[14] The new higher prices took into account recent wages awards, the high price of sugar on the international market, and the need for the industry to achieve a rate of growth sufficient to service the expanding domestic demand for sugar.

Beyond these moves, the only significant variation in the nature of state control of the Australian sugar industry over the next decade, was that instituted by Prime Minister Bruce in 1923, when he handed the financial and marketing control of the industry over to the Queensland Government provided that the price of raw and refined sugars were reduced in line with international trends.[15] Under this arrangement, the Queensland Government established the Queensland Sugar Board to administer the marketing of the commodity. To all intents and purposes, however, the marketing and distribution structures remained unchanged. The CSR still acted as the state's Agent, but now, it reported to the Queensland rather than the

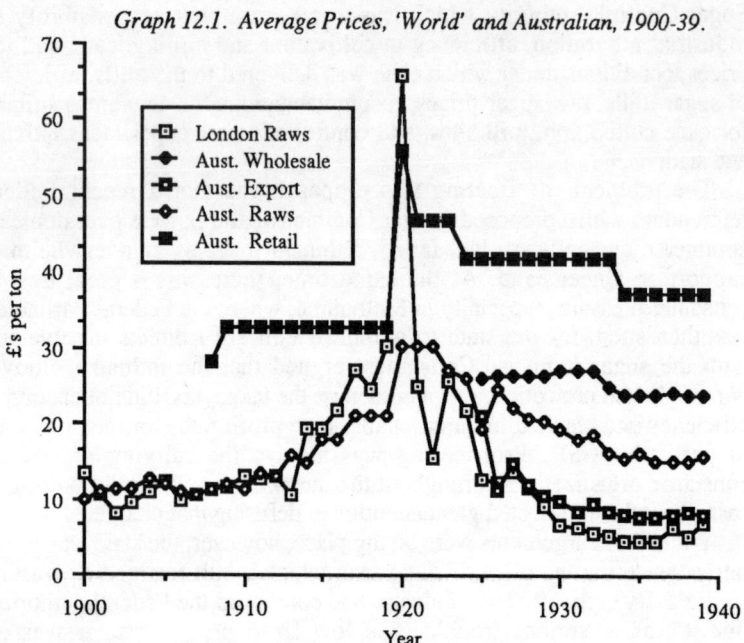

Graph 12.1. *Average Prices, 'World' and Australian, 1900-39.*

Sources: All Average Prices. London Raws; Lowndes, A.G., *South Pacific Enterprise*, (Sydney, 1956), 443. Australian Wholesale Price; Lowndes, 447. Australian Export Price; Reports of the Sugar Experiment Stations, published annually in the *Queensland Parliamentary Papers* Australian Raws; Easterby, 141, Lowndes, 447. Australian Retail; Lowndes, 447.

Federal Government. In the meantime, Federal import controls on sugar and allied food products remained in force, subsidies were directed into some of the allied industries, and State support for the industry at the scientific and administrative levels continued with enthusiasm. The State also continued to encourage the settlement of sugar lands under favourable conditions, and it was during the interwar period that soldier–settler schemes saw returned servicemen take up cane farming.

By contrast with the war–time period, protection of the industry and stringent price controls over cane and sugar products created extremely favourable conditions for the expansion of cane sugar production in the 1920s and marked improvements in productivity and efficiency.[16] First, the tonnage of cane reaped per acre improved steadily between 1923 and 1928, and the violent fluctuations in cane yields evident in preceding years steadied down. Certainly, climatic conditions were more favourable in the twenties than in the preceding decade, but there is evidence that improved cane varieties, better cultivation methods, including the more scientific application of both green and artificial fertilizers, irrigation and drainage schemes, mechanization of field work, planting inter–row cultivation, weeding and fertilizer distribution, and improved transport from field to factory also contributed to the better yields.[17] The intensity of mechanization, however, varied markedly from region to region in Queensland. Annual average expenditure on oils and grease in the northern districts, for example was more than double that in the southerly region.[18] Farms in the north tended to be larger and much more productive than the southerly farms. Northern farmers were not only in a better financial position to invest in tractors therefore, they had a greater incentive to innovate at the level of cultivation technology since labour costs in the north were significantly higher than in the south.

Perhaps the most startling technological innovations were applied in the mills. Mill capacity was not only increased markedly but the extractive and manufacturing processes were also improved.[19] In 1929, Queensland claimed the world record for producing one ton of sugar at 94 per cent net titre from the least amount of cane at 6.9 tons of cane. While the technology of mills was improved on the advice of international experts, visitors to the Queensland industry in the late twenties and early thirties marvelled at the locally developed 'gadgets', which had been applied both to field and factory work in the quest for greater efficiency.[20] The Bureau of Sugar Experiment Stations, enjoyed great success in the twenties in its scientific work on the cultivation side of sugar production and it was during this period that the technical staff of the Bureau established the Queensland Society of Sugar Technologists, which was affiliated to the international society. The result of the industry's expansion and improved efficiency, was that it quickly satisfied the domestic demand for sugar and its exports began to grow rapidly.[21]

Clearly, the favourable price regime which guaranteed Australian producers prices well above world prices, as well as a monopoly access to the growing domestic market, provided an incentive to the industry not merely to expand and improve efficiency, but a rate of capital accumulation sufficient to quickly achieve those objectives. Other factors included an especially attractive export price for sugar in 1926, which stimulated the planting of virgin lands which naturally produced heavy crops; the encouragement of the idea that British Preference on Dominion sugars would be materially increased, a prospect suggested to the industry by the President of the British Empire Sugar Producers' Association during his visit to Australia in 1928; the 'propaganda' that was spread throughout the sugar districts about the same time, that surplus sugar could be profitably disposed of in the production of power alcohol, and that further revenue would accrue from the use of megass in the production of celotex, paper and artificial silk.[22] By the end of the decade, however, the Australian

Table 12.1. Queensland Sugar Production, 1900-1939.

Year	1	2	3	4	5	6	7	8	9	10	11
1900	108535	72651	848328	92554	1.274	11.677	9.166	58	1596	90.40	669389
1901	112031	78160	1180091	120858	1.546	15.098	9.764	52	2324	104.10	789191
1902	85338	59102	641927	76626	1.297	10.861	8.377	43	1782	3.60	934932
1903	111516	60375	823875	91828	1.521	13.646	8.972	39	2355	2.40	646875
1904	120317	82741	1326989	147688	1.785	16.038	8.985	53	2787	3.00	1257615
1905	134107	96093	1415745	152722	1.589	14.733	9.270	51	2995	11.30	1448845
1906	133284	98194	1728780	184377	1.878	17.606	9.376	52	3546	9.40	1615689
1907	126810	94384	1665028	188307	1.995	17.641	8.842	52	3621	18.60	1779643
1908	123902	92219	1433315	151098	1.638	15.543	9.486	50	3022	15.00	1482382
1909	128178	80095	1163569	134584	1.680	14.527	8.646	48	2804	8.20	1125887
1910	141779	94641	1840047	210756	2.227	19.447	8.733	51	4132	6.70	404
1911	130376	95766	1534451	173296	1.810	16.023	8.855	51	3398	7.40	10001
1912	147562	78142	994212	113060	1.447	12.723	8.794	48	2355	2.30	1051
1913	147743	102803	2085588	242837	2.362	20.287	8.588	49	4956	3.10	78
1914	161195	108013	1922633	225847	2.091	17.800	8.513	46	4910	18.70	54
1915	153027	94459	1152516	140496	1.487	12.201	8.203	45	3122	1.10	1798
1916	167221	75914	1579514	176973	2.331	20.807	8.925	43	4116	1.00	112
1917	175762	108707	2704211	307714	2.831	24.876	8.788	46	6689	2.10	107
1918	160534	111572	1674829	189978	1.703	15.011	8.816	42	4523	2.10	198
1919	148469	84877	1258760	162136	1.910	14.830	7.764	32	5067	2.90	336
1920	162619	89142	1339455	167401	1.878	15.026	8.001	34	4924	4.30	721
1921	184513	122956	2287416	282198	2.295	18.604	8.106	40	7055	2.00	77

cont.

Year	Col 1	Col 2	Col 3	Col 4	Col 5	Col 6	Col 7	Col 8	Col 9	Col 10	Col 11
1923	219965	138742	2045808	269175	1.940	14.745	7.600	37	7275	15.80	116
1924	253519	167649	3171341	409136	2.440	18.917	7.751	37	11058	84.10	149891
1925	269509	189466	3668252	485585	2.563	19.361	7.554	37	13124	212.20	962736
1926	266519	189312	2925662	389272	2.056	15.454	7.516	36	10813	117.70	2206435
1927	274838	203748	3555827	485745	2.384	17.452	7.320	35	13878	130.00	941118
1928	283476	215674	3736311	520620	2.414	17.324	7.177	35	14875	211.10	1848059
1929	291660	214880	3581265	518516	2.413	16.666	6.907	35	14815	196.30	2063091
1930	296070	222044	3528660	516783	2.327	15.892	6.828	35	14765	191.30	2067232
1931	309818	233304	4034300	581276	2.491	17.292	6.940	35	16608	277.30	1934419
1932	291136	205046	3546370	514027	2.507	17.295	6.899	33	15577	222.20	3128208
1933	311910	228154	4667122	638559	2.799	20.456	7.309	33	19350	271.00	1792568
1934	303926	218426	4271380	611161	2.798	19.555	6.989	33	18520	551.80	2837712
1935	314700	228515	4220435	610080	2.670	18.469	6.918	33	18487	277.70	2715833
1936	338686	245918	5170571	744676	3.028	21.026	6.943	33	22566	409.50	2739990
1937	348840	245131	5132886	763242	3.114	20.939	6.725	33	23129	436.80	3692519
1938	347199	251847	5342193	778064	3.089	21.212	6.866	33	23578	476.00	4007844
1939	353996	262181	6038821	891738	3.401	23.033	6.772	33	27022	386.60	4155785

Notes: Col 1: Acres under Cane; Col 2: Area Crushed; Col 3: Weight of Cane Crushed (tons); Col 4: Sugar Manufactured (tons) (after 1904, 94% titre); Col 5: Average Yield, Sugar manfactured per acre of cane crushed (tons per acre); Col 6: Average Yield, tons of Cane per Acre (tons per acre); Col 7: Average Yield, Area of Cane Crushed per Ton of Sugar Manufactured (tons per Acre); Col 8: No. of Mills; Col 9: Sugar manufactured per mill; Col 10: Total Australian Sugar Exports ('000s of tonnes) (between 1910 and 1923, six months only); Col 11: Export Value (A£'s);

Sources: Cols 2,3,4,8,11; drawn from tables published annually in *Statistics of Queensland*; Col 10, FAO, *The World Sugar Economy in Figures*, (Rome, c.1960); Cols 5,6,7,9, calculated on the basis of data in the Table.

industry was now plunged into a major crisis of over–supply. Instead of complaining about low prices on the domestic market and seeking access to a lucrative export market, it now began to carp about the poor returns its abundant, surplus product received abroad. The only thing that had not changed was its appeal to the state to solve its problems.

The 1930s

In June 1929, the Queensland Government convened a conference of sugar growers, millers, refiners in an attempt to sort out the crisis of over–production. The solution which they adopted was termed the 'Peak Year Scheme'. Under this system, the future output of each mill in Queensland was limited to its highest output in any year since 1915. Sugar produced outside of the quota was placed in a separate export pool and valued at the lower export price. The Queensland Sugar Board was the authority which fixed the peak years according to the resolutions of the conference. Under the scheme, mill output was limited to 611,608 tons. it should perhaps be stressed, that the Peak Years Scheme was voluntary and put into effect by the industry itself. There is some suggestion that the industry embarked on this strategy to avoid the inevitability that the state would undertake the task of curtailing production for it. In fact, since 1925, the Queensland Cane Prices Board had been attempting to do just that.

The Board announced in March 1925, that it would not assign any new cane land[23], nor would the land already assigned in the name of specific growers be increased. The policy hardened in 1927, when a procedure was promulgated to delimit the areas already assigned to growers 'having a prescriptive right to assignment of any portion of their land as cane growing lands'. Indeed, the general assignments of all cane growing land were revoked and re–assignment was made in such a way as to exclude altogether virgin land, land unprepared for cane cultivation, or land which had been out of cultivation for three years previously. Land assigned included only those portions in current use for cane growing, including fallow or land in preparation for planting. By 1931, the delimitation of land by the Central Cane Prices Board was virtually complete.[24] Certainly, the rate of growth of the area in cane slowed down considerably after 1925.[25] It is extremely difficult, however, to calculate the extent to which this was due to the delimitation of assigned land, or to the cane farmers' responses to adverse conditions in the international economy. There are some suggestions that the delimitation exercise developed less as a strategy to cope with the crisis of over–production than as an attempt to control the number of 'aliens' taking up sugar cultivation. The involvement of southern European farmers in the Queensland industry became a major issue in the inter–war period.

Sugar workers and small cane farmers, mainly soldier–settlers, appear to have been the major source of opposition, which came to a head in 1926 and 27, to the mainly Italian and Maltese immigrants. The Italians, it was argued, threatened wage levels and living standards and their tendency to purchase land cooperatively forced land prices up and out of the reach of the small 'British' farmer. These objections were invariably expressed in nationalist and virulently racist terms echoing arguments made two decades before that the immigrants threatened the preservation of the White Australian Policy and the settlement of the north for defence purposes.[26] Despite the relatively low percentage of Italians in the sugar districts, however, the Commonwealth Government bowed to political pressure and sharply reduced the number of visas granted to southern Europeans. Assignments of land under the Cane Prices Boards to farmers new to the industry also fell. And the Australian Workers' Union came to agreements with northern mills limiting the number of southern Europeans employed in the industry and giving preference to a large number of British gangs. Subsequently, the number of Italians employed in the industry fell sharply.[27]

The third avenue by which the state sought to control sugar output during the thirties, was through its pricing and export policies. Although the various domestic sugar agreements guaranteed producers a minimum return on their output destined for home consumption, the fixed price trend was downwards over the next decade more or less in line with the international price movements. It was true, the domestic price of sugar was always significantly higher than the international price, but costs of production increased during the thirties. This, combined with the operation of the Peak Year Scheme, restrictions over cane land assignments, and the very low returns growers received for surplus, exported sugar, tended to moderate the rate of growth of sugar production over the period. Interestingly, unspectacular growth rates were not inconsistent with continued improvements in productive efficiency.[28]

By the end of the thirties, domestic demand had improved but slightly, partly through population growth, by an increasing consumption of sugar per head, and the success with which the industry had diversified its products. At the same time, as income rises and becomes more evenly distributed, the consumption of sugar becomes relatively more inelastic. These conditions obtained in Australia by the end of the 1930s. A brighter prospect on the horizon, however, was the very satisfactory quota which Australia, following intense diplomatic and industry lobbying had been awarded under the International Sugar Agreement of 1937 of 400,000 long tons, and the promise that this would be improved under the Empire Preference Scheme. This followed increasing exports of Australian sugar during the 1930s.[29] Under these circumstances, the industry urged continued state control at the end of the decade, only with some improvements to the established pricing and land assignment mechanisms in order to iron out disparities in the returns to some mills and districts. Gentle encouragement was called for, and the state gave every indication that it was willing to go on playing the supportive role.[30] Only the outbreak of the war in 1939 turned the world upside down again for the sugar industry in Australia.

Conclusion

Between 1900 and 1939, The State of Queensland and the Government of the Commonwealth carried out a dozen major Inquiries, including six Royal Commissions into the Queensland Sugar Industry, and several others which bore upon the industry's affairs. Both the State and Federal legislatures enacted numerous pieces of legislation, promulgated associated regulations, set up a dazzling variety of bodies to administer the affairs of the industry, devoted considerable sums of money to fund the scientific investigation of cultivation and manufacturing processes, subsidized and protected the industry from outside competition including the use of embargoes, tariffs and bounties, and encouraged the development of allied manufacturing industries. Not for the first time, the Federal Government even manipulated immigration controls to placate sugar interests. It is clear that the state in Australia, Federal or Provincial, of Conservative or Labor hue, was committed not merely to the survival of the Australian sugar industry, but to its robust development, regardless of the prevailing political or economic conditions.

The benefits of state support appear to have been spread very evenly through out the industry, flowing to cane farmers, millers, refiners and distributors and, considering the level of wage rates, even to sugar workers. One of the most important functions of the state was in its role as an arbitrator between the various sections of the industry which were naturally in conflict. Creating conditions of relative harmony between the competing sections of the industry, and ensuring its stability under the conditions of crisis in the international sugar economy, not to mention the effects of the Great Depression, must be regarded as a major achievement. Certainly, state intervention was materially, highly beneficial for the Queensland sugar industry and its allied manufacturing industries. The growth of the industry was the envy of its competitors, and its achievements in improving productivity were very impressive. The irony is, that it took an armoury of controls first devised to cope with the crisis of wartime conditions (many of which remain in place to this day) to achieve the triumph. The historical experience of the Queensland industry in the inter-war years confronts the view popular in some economic and political circles that state control, protection and subsidies militate against the development of efficient, expansive industries.

Nevertheless, state involvement in the Queensland sugar industry was not popular with consumers or the Australian electorate at large This potentially volatile opposition was handled very skilfully by both the state and the industry, through the appeal to a range of very powerful arguments to justify the level of state support from which the industry benefited. Although the specific content of the arguments changed with the circumstances, state policy on the sugar industry was persistently justified on nationalist and racist grounds. If in 1901, sugar producers believed that the introduction of the White Australia Policy would lead to their ruin, subsequent events would prove both the ideology of the Policy and its practise, to be the industry's saviour. The only variation on this theme occurred in the aftermath of Great Depression, when the thoughts of John Maynard Keynes emerged as the compelling justification of state control and protection of Australian cane sugar production.

References

1. See Table 12.1 for the industry's output, efficiency and export figures for the period under examination.

2. In this essay, 'State' refers to the Provincial political entity and 'state' is the more general term. It should also be pointed out that although cane sugar production was pursued in northern New South Wales throughout the period under examination, Queensland sugar production accounted for the vast bulk of Australian sugar output. A small beet sugar industry developed in Victoria in this period, but was of little consequence in the sugar industry taken as a whole.

3. For a detailed analysis of the economic history of the Queensland sugar industry in this early period and full references to the literature see Adrian Graves, *The Political Economy of the Queensland Sugar Industry, 1862–1906*, (London, The Royal Historical Society, forthcoming).

4. See Table 12.1, Cols 1, 2, 3 & 4.

5. *Commonwealth Parliamentary Papers*, 1912, 1042. See also 1055, for a very clear statement of this almost unique ideology of industry protection.

6. Easterby, H.T., *The Queensland Sugar Industry. An Historical Review*, (Brisbane, 1933), 112.

7. Called the *Regulation of Sugar Cane Prices Acts*.

8. Easterby, 143–4.

9. *Report of the Royal Commission on the Sugar Industry, 1920*, (Canberra, Government of the Commonwealth of Australia, 1920), xxxiv–v.

10. *Ibid.*

11. CSR, 'The Australian Sugar Industry. A Sort of History'. (Sydney, 1974, typescript), 23.

12. *Reports of the Sugar Inquiry Committee of 1931*, (Canberra, Government of the Commonwealth of Australia, 1931), 35. Exports of jams, for example, rose from 2 million lbs just before the war to 79 million lbs by 1919. CSR, 'The Australian Sugar Industry', 21.

13. *Report of the Royal Commission on the Sugar Industry, 1920*, (Canberra, Government of the Commonwealth of Australia, 1920), xlvii–xlviii.

14. See Graph 12.1.

15. To observe how this operated in the various sugar Agreements over the decade, see Graph 12.1.

16. See Table 12.1, Cols. 5, 6, & 7.

17. *Reports*,...1931, 69.

18. See the evidence of the Queensland Cane Prices Board, *ibid.*, 111.

19. See Table 1, Col. 9.

20. *Reports*...1931, 1031–92.

21. See Table 12.1, Cols, 10 & 11.

22. *Reports*...1931, 59.

23. Under the provisions of the *Sugar Cane Prices Acts, 1915–22*.

24. Reports...1931, 90.

25. This can be observed by reference to Table 12.1, Col. 2.

26. *Reports*...1931, 10.

27. Easterby, pp. 169–70.

28. See Table 12.1, Cols 5, 6 & 7.

29. See Table 12.1, Cols.10 & 11 .

30. 'Report of the Royal Commission on Sugar Peaks and Cognate Matters, 1939', *Queensland Parliamentary Papers*, 1939, II, 1031–92.

13

Treacherous Cane: The Java Sugar Industry Between 1914 and 1940

Peter Boomgaard

Introduction

Between 1840 and 1940 the Javanese sugar industry was responsible for between 5 to 10 per cent of the world's total sugar production. It was during the period 1914 to 1939, however, that Java's sugar experienced both its finest hour and its deepest gloom. In 1928 Java produced almost 12 per cent of total global output, only to reach an all time low of 2 per cent in 1935 and 1936 (Table 13.1). Of course, a drop in production was to be expected during a world–wide depression, but not all sugar producing countries experienced such a drastic reduction in their share of global output, nor such a peak just before the onset of the 1929 crisis. Nor was such a decline typical for all exports from the Netherlands Indies. For example, the shares of cinchona, kapok, pepper, rubber, and tea were fairly stable or even increased between 1930 and 1940.[1] Finally, it was not the first time that Java's sugar industry had been confronted with adverse conditions, but never before had it been beaten so badly.

In this chapter I shall trace the internal and external causes of the industry's varying fortunes between the two world wars. However, because the external causes (international trade policies) are discussed in the introduction to this collection, I will concentrate mainly on the characteristics which typified the sugar industry of Java.

The Java sugar industry

Table 13.1. Javanese Sugar Production 1914-1940.

Year	1	2	3	4	5	6	7	8
1914	186	147	15132	1405	183	7.4	27.2	–
1915	186	151	14412	1319	213	7.2	27.6	–
1916	186	157	16246	1630	259	9.7	29.9	–
1917	185	160	17353	1822	212	10.6	27.2	–
1918	186	163	15882	1778	184	10.5	27.2	–
1919	179	138	13284	1336	762	8.3	35.5	–
1920	183	159	14628	1544	1049	10.1	46.5	–
1921	184	159	15169	1685	415	10.0	34.8	74550
1922	182	161	17025	1808	271	10.1	23.7	80455
1923	180	163	16334	1793	498	10.0	36.4	77287
1924	181	172	18318	1998	488	10.1	31.6	77566
1925	179	178	19328	2300	365	9.8	20.3	77968
1926	178	180	18982	1973	265	8.2	16.7	77484
1927	176	186	21451	2379	359	10.2	21.9	81054
1928	178	195	25700	2948	369	11.7	23.4	87439
1929	179	197	24527	2898	304	10.8	21.1	88512
1930	179	199	25698	2971	244	11.1	21.1	86196
1931	178	201	26518	2839	125	10.2	16.7	81418
1932	165	166	22393	2560	97	10.2	17.9	72803
1933	97	84	11531	1372	61	6.0	13.0	41958
1934	50	34	5148	636	45	2.6	9.2	21523
1935	40	28	4114	510	35	2.1	7.8	17945
1936	37	36	4723	575	34	2.1	6.3	17147
1937	81	84	12084	1380	50	4.8	5.3	37168
1938	80	85	11875	1376	44	4.7	6.7	37584
1939	84	95	13091	1562	77	5.5	10.3	39912
1940	85	91	12426	1587	52	5.4	5.9	40893

Note: Col 1: Number of Active Mills; Col 2: Area under cane in '000s of Hectares; Col 3: Cane production in '000s of Tons; Col 4: Sugar production in '000s of Tons; Col 5: Sugar exports in '000,000s of NI Guilders ; Col 6: Java sugar output as a percentage of world output; Col 7: Java sugar exports as percentage of Netherlands Indies' export (value); Col 8: Number of permanent workers.

Source: *Changing Economy in Indonesia* , Vol.I, (P.Creutzberg), Tables 1 and 6; *Changing Economy in Indonesia* Vol.VIII, (W.Segers), Table 12; H.E.Levert, *Inheemsche Arbeid in de Java Suikerindustrie*, Wageningen, 1934, 331; M. Moreno Fraginals, *El ingenio; compléjo económico social cubano del azúcar*, Havana, 1978,(III Vols.), III, 35-40; *Statistisch Jar-Overzicht van Nederlandsch-Indië*, 1940, 294.

Land and labour

In 1900 almost all cane was grown on land hired by the sugar factories from the local indigenous population. All work was carried out by wage–labour. If we accept that there are basically two systems of cane–growing in any tropical agrarian society, cane as a plantation crop or as a peasant crop, the system of production in Java at that time must be regarded as a plantation system, even though the factories did not own their fields. It was the manager of the sugar factory who decided what to plant when and where, who hired the workers, and who took the responsibility for all production decisions.

This system of production represented a fairly recent stage in the development of the sugar industry. Cane had been grown in Java before the Dutch arrived in about 1600. The estate production of sugar in the environs of Batavia had become prosperous by about 1700, but 50 years later it started to decline. This was partly due to erratic policies of the Dutch East India Company and partly to problems with credit, firewood and labour. Around 1800 the St. Dominque slave rebellion and its aftermath stimulated a short–lived sugar boom in Java[2], but it was not until 1830 that Java's sugar production started to expand regularly. This was largely caused by the so–called Cultivation System (*Cultuurstelsel*) under which the peasantry had to plant sugar (and other crops for the European market) on part of its arable lands. Processing, however, was in the hands of European and Chinese (and very seldom Javanese) sugarmillers, who had obtained contracts from the Netherlands Indies, Government. Sugar export was largely a Government monopoly, for which it contracted with the NHM (*Nederlandsche Handel–Maatschappij*), a trading company sponsored by the Government of the Netherlands. The sugar was finally sold in Holland, where the returns were used to replenish the empty treasury. So cane production was in the hands of the peasantry, processing was done by private Western enterprise, and the sugar trade was a government affair.[3]

After 1870 this system was gradually abolished, because in the Netherlands the political tide had turned. To the Liberals, now in power, government interference with export production was anathema. Compulsory cane planting was reduced annually with a fixed percentage over a period of 21 years, and the export monopoly came to an end. New regulations were drawn up for leasing village land and hiring labour, and from then on every sugar factory had to obtain its own land and labour.[4]

So at the beginning of the twentieth century sugar was grown as a plantation crop on lands leased from the local population. All activities in the fields and in the factory were carried out by wage–labour. Expansion of the cane fields was possible, but surrounded with restrictions, and subject to costly obligations.

Capital

Until 1800 not much capital was needed to run a sugar–mill. However, with the construction of modern water–mills the sugar industry became more capital intensive. During the first ten years of the Cultivation System the Netherlands Indies Government had to advance money to the millers in order to get the industry started. Soon, however, it became clear that the sugar–contracts were highly profitable, and both the King, the government in the Hague, and the Netherland Indies' Government were swamped with requests for sugar contracts. The sugarmillers were, indeed, in a very comfortable position in that they bought cane and sold sugar for a fixed price. It was the peasant cultivator who ran the risk of a harvest failure and the government that had to face price fluctuations on the world market.

Much of this changed after 1870, and even more after 1884/5 when the Javanese sugar industry was hit simultaneously by the international sugar crisis and the so–called sereh–disease. The sugarmillers were confronted with a sudden price fall, harvest failures, and nervous bankers who tried to call in their loans. A major crisis was avoided only because the banks were backed up with huge loans from the Amsterdam capital market, and many sugar factories were taken over by banks. Until then many factories has been owned by individual entrepreneurs, but after the crisis most enterprises were converted to limited liability companies with headquarters, and most of the shareholders, in Holland. Many small firms were liquidated. Now sufficient capital was available for a thorough modernization of the factories, which was long overdue.[5] Only if costs per unit of product were lowered, and quality improved, could competition with other cane producers and the protected beet sugar industry be kept up. Modernization of the factories was one answer to this problem. Higher yields per acre was another. Stimulated by the rapid spread of the sereh–disease, a number of experimental stations were set up between 1885 and 1887. Although they failed to discover the cause of the sereh–disease, the experimental stations found ways to get around it. They were also instrumental in the development of high yielding varieties.[6]

So the Java sugar industry entered the twentieth century highly capitalized and thoroughly modernized, backed up by advanced agricultural research.

Markets

Up to 1870 almost all Java sugar had been shipped to the Netherlands. After 1872 the government sales of sugar stopped, and exports were now directed towards London, the world's most important sugar market. Very soon, however, the competition from beet sugar made the European markets unattractive for cane, and new outlets for Java sugar had to be found. It was Java's good luck that just at that moment Cuba saw its production curtailed by internal problems (abolition of slavery, the ten-years war with Spain), which created a shortage in the United States. So for a period of some twenty years America was an important buyer of Java sugar. After 1900, however, this market became virtually inaccessible, because after the Spanish–American War Cuban sugar received preferential treatment in the US.

Exporting to Asia had started around 1880, and with Europe and America gone, Asia became Java's most important buyer after the turn of the century. Soon 80–90 per cent was going to Asia: Singapore, China (and Hong Kong), Japan (and Formosa), British–India, and Australia. Until 1930 these markets were responsible for 75 per cent, on average, of Java's sugar exports.

Although the Brussels Convention(1902) had helped to curb beet sugar protection somewhat, thereby improving the position of cane, Java made its come–back on the European market only after the outbreak of World War I, when exports from Germany, Austria and Russia were curtailed. During the war years somewhat more than one third of Java's sugar was sent to Europe. However, when the European beet sugar industries began to recover after 1920, Java was, once again, forced to depend on Asian markets.[7]

But dark clouds were gathering. Although the Brussels Convention and WWI had been instrumental in keeping prices from falling and world production from increasing too much, by 1924 it became clear that over–production was a fact. When prices dropped still further in 1929 and 1930, the so–called Chadbourne Plan led to an international agreement in 1931. Java participated, Holland did not. Nor did a number of other countries. As the non–participants increased their (protected) production, stocks kept piling up in the exporting countries, and prices kept falling.

Simultaneously Java was rapidly losing its Asian markets because of tariff barriers in British–India, Japan (with Formosa) and China, erected to protect the existing or newly developed sugar industries. Although the Netherlands and Britain also protected their markets, exports to Europe increased slightly. Nevertheless Java, for the first time in a century, had to carry out a drastic reduction of its area under sugar, because it could not find new outlets. After the London Sugar Convention of 1937 prospects became a bit brighter, but even in 1940 production had not yet reached the level of 1921 (Table 13.1).

Getting the producers organized

The 1884/5 crisis had led to more co–operation between Java's sugar producers. The establishment of the experimental stations between 1885 and 1887 had been the first important result. In 1894 government proposals for export duties on sugar and new regulations on hiring land acted as a catalyst on the already existing tendency to present a united front, and the General Syndicate of Sugar Producers in the Netherlands Indies (the Syndicate), was established. It was an organization of the managers of the sugar factories, residing in Java. Most factories, however, by then mostly limited liability companies, had their headquarters and, therefore, their owners in the Netherlands. Until 1917 this state of affairs did not constitute a problem, but when in that year the managers failed to come up with a solution for the export problems caused by the war, the owners took over, and in 1917 the Association of Owners of Netherlands Indies' Sugar Enterprises (BENISO) was established in Holland. They decided that the situation called for a sellers' cartel of all producers. The result was the United Java Sugar Producers (VJSP),with a representative in Surabaya, Java (1918). The VJSP was highly successful in selling most of the sugar stock that had been piling up in Java, and in 1919 the situation was back to normal.[8]

Soon, however, other threats had to be faced. The years 1918–1921 were characterized by social and political unrest in the Netherlands Indies, and by government attempts to meet the demands of the indigenous political leaders by reform measures. Both unrest and reform proposals were directed partly against the sugar industry, which was seen as the symbol of western capitalism, with its large profits that were remitted to Holland, and its refusal to give their (Javanese) workers an adequate share in its wealth.

In 1921 the Syndicate decided to establish a separate Java Sugar Employers Union (JSWB), in order to harmonize the industry's labour relations and labour regulations, and to present a united front to trade unions and government. During the first years of its existence its role was almost entirely defensive, but from 1926 onward it had taken the initiative in a number of improvements of labour relations (e.g. pensions). In 1931 the Netherlands Indies' Government issued a number of sugar–export regulations, in order to implement the Chadbourne Agreement. Because the VJSP was unable to arbitrate between its members regarding the division of the export quota, the government stepped in with a new series of decrees, which led to the establishment in 1932 of the Netherlands Indies' Association for the Sale of Sugar (NIVAS). This organization was granted a sugar–export monopoly. On its board of directors were represented the producers or managers, the Governor–General, and the Javasche Bank (Central Bank of Netherlands Indies). With the founding of the NIVAS as single–seller, there was no need any more for a sellers' cartel, and the VJSP was disbanded.[9] Between them, the Syndicate as an organization of the managers, and the NIVAS with its export monopoly, controlled the entire industry. For the third time in its history Java's sugar industry was operating under monopoly conditions.

The workers between protests and pensions

From its very beginning as a government sponsored undertaking in the 1830s the sugar industry has been both praised as the mainstay of Java's economy, and branded as the main cause of the poverty of its indigenous population. After 1890 the industry was supposed to rely entirely on wage–labour and village land, obtained voluntarily from the villagers. In practice, many a sugar–miller restored to various measures of coercion in order to be assured of an adequate supply of land and labour. Even as late as 1915 the Sarekat Islam leaders cited many instances of sugar factories that used their influence with the Dutch and Javanese officials in order to obtain land from unruly villagers.[10]

The sugar industry, although highly mechanized after the huge capital influx in the 1880s, needed a large labour force. Only the factory itself was mechanized, and all agricultural labour was done by hand. It could even be argued that labour requirements per plant had increased, because more cane was needed in order to use the new machines to their full capacity. Only then could this expensive machinery be exploited at a profit. The labour problem was two–fold. It was hard to get sufficient labour, especially during the harvest campaign (May–September), but it was still harder to make the seasonal labourers show up regularly. This disciplinary problem was related to the structure of the Javanese economy at that time. Seasonal work in the factory or in the cane fields was a side–line occupation for a still predominantly peasant labour force. Although the majority of this work force needed all the income from the factory they could get in order to survive, a sizable proportion of the casual labourers came there with a target–income in mind. They wanted to work off a debt, or needed a specific amount of money for one of the many life cycle ceremonies that Javanese custom prescribed. When the required amount had been earned, they simply vanished.

163

Between 1920 and 1930 for 180 factories 150,000 to 175,000 factory-workers were needed, of which two thirds were wanted during the campaign only. For the cultivation of cane some 800,000 to one million seasonal labourers were required. This did not include cutting and carting, for which an additional 250,000 people had to be found. Between 12 and 14 per cent of the factory workers were women, and another 2 to 5 per cent children. Of the total labour time spent in the cane fields about 30 per cent was done by women and 15 per cent by children.[11]

After the WWI shortage of labour ceased to be a major problem, but now the managers were confronted with a sudden outburst of labour unrest. Before the war labour protest was largely confined to walk–outs. Disgruntled workers just left, and other people were hired in their place. Strikes, defined as refusal to work in order to obtain better conditions, were virtually unknown. Grievances against the sugar factories found a traditional outlet in cane–burning, of which the incidence can be used as a rough index of frustration. After the war the incidence of cane–burning tended to decrease, partly because the sugar factories were successful in curbing it with effective counter measures, partly because social unrest found a new outlet in the Nationalist movement, which was able to capitalize on the existing 'industrial' frictions.

In 1920, a wave of local strikes broke out, organized by the Union of Factory Personnel (PFB); about 25,000 workers participated in 72 strikes directed against 65 factories. It was estimated that almost 86,000 work-days were lost, so the average duration of the strikes must have been rather low. In 1921 about 1,500 workers took part in 17 strikes, but after this year the PFB rapidly lost ground, and strikes became rare. Between 1919 and 1921 the general wage level in the industry increased, no doubt partly as a result of the strikes. Another result was the creation of the JSBW (the employers union), mentioned in the preceding section.

After the disappearance of the PFB, and the failure of the left–wing unions to organize the sugar workers, traditional patterns of protest reasserted themselves. In 1919 and 1920, when the PFB had been active and attractive, the incidence of cane–burning fell, only to increase again in 1921, when it became clear that the PFB was not a real alternative. During the period of Communist inspired agitation (1924/6) the graph of cane-burning shows another peak.[12] I have not been able to locate data on cane-burning and strikes after 1932, and they are not mentioned in the more general literature. It is likely that very little of it did occur after this year, given the fact that there was less cane to be burned, and every cent to be earned with casual labour was most welcome in this period of low wages.

Is it one of the most fascinating aspects of the Javanese sugar industry that it embodies so clearly the 'dual' aspects of a colonial economy and society. On the one hand it was the focus of 'primitive' peasant protests like cane–burning, on the other it was also the cradle of 'modern' attitudes towards labour relations. For its permanent labour force the industry provided model villages with modern sanitation, hospitals and free medical care, pension and gratuity plans, and free schooling. Of course this was a very small island in a sea of casual labour. The total wage–bill of the sugar industry in 1924 equalled the sum of all wages paid out by all other important enterprises together. In 1925, 37.3 per cent of total production costs had to be spent on labour, of which 7.3 per cent was paid to European personnel, and therefore 30 per cent to indigenous labour.[13]

After 1932 the amount of labour needed by the sugar industry dropped sharply. So did the wages. Although those workers who were still employed did not experience a fall in real income, because prices dropped at least as much as wages, it is clear that the mass of unemployed workers, both permanent and casual, had to face hard times. Because the area under cane was also reduced considerably (Table 13.1), many villages lost both their income from labour and from land. A massive return to subsistence farming took place. Hardest hit were those unemployed factory labourers who had relied entirely on the factory for their sustenance. Many had cut themselves loose from the village sphere, and had nowhere to go after their unemployment benefit had been spent. Once the pride of the colonial 'dual' economy,they were now its most lamentable victims.

High yielding varieties (HYVs).

In 1931, when the area under cane reached its maximum, sugar production was twice as high as it had been in 1914, although the area under cane had expanded only one third (Table 13.1). This implies that the average sugar yield per hectare had increased considerably. Most of this increase (80 per cent) can be explained by a higher yield of cane per hectare (Table 13.2). The most striking jump in yields per hectare occurred between 1922/7 and 1927/32. How is this sudden increase to be explained?

The answer is by the 'invention' and rapid spread of POJ 2878. This somewhat mysterious code stands for experimental variety no. 2878, developed by the Experimental Station of Eastern Java (*Proefstation Oost Java*, or POJ for short). Due to the fact that the experimental stations had succeeded in convincing the factories that they needed experimental plots to be able to test new varieties in a scientific way, POJ 2878 could spread as rapidly as it did.[14]

The Java sugar industry

Table 13.2. Productivity in the Javanese Sugar Industry 1914–1940.

Index:1921/4=100	1914/17	1921/4	1929/32	1937/40
sugar/ha	90	100	133	150
cane/ha	100	100	127	137
sugar/1,000 cane	89	100	104	109

Source: *Changing Economy in Indonesia* , Vol.I, ((P.Creutzberg), Tables 1 and 6; *Changing Economy in Indonesia* Vol.VIII, (W.Segers), Table 12; H.E. Levert, *Inheemsche Arbeid in de Java Suikerindustrie,* Wageningen, 1934, 331; M. Moreno Fraginals, *El ingenio; compléjo económico social cubano del azúcar,* Havana, 1978, (III Vols.), III, 35–40; *Statistisch Jar–Overzicht van Nederlandsch–Indië,* 1940, 294.

So when the crisis struck, Java's sugar yield per hectare, already in the top bracket of the sugar producing countries, had just increased considerably. In 1923 Java, Hawaii, and Peru produced about 11 tons per hectare. This was twice as much as the average yield of the next group of producers, the Philippines, Brazil, Surinam, and British Guiana.[15] In 1930 Java produced 15 tons/hectare. I could find no figures for other countries around the same time, but it is not likely that higher yields were obtained elsewhere. In 1940 the average yield was 17.4 tons/hectare, but this was due more to the fact that the industry had reduced its area under cane, no doubt getting rid of the less productive lands first, than to the spread of new HYVs, although some new, and presumably better POJ varieties had been introduced after 1933. It is not surprising that foreign cane producers were avid readers of the experimental stations' publications. It can be said that their sphere of influence went far beyond the shores of Java.[16]

Production and productivity, prices and profit

In the final analysis the question must be asked whether POJ 2878 has been a blessing or a curse. It enabled the industry to increase its share in a highly competitive market where prices were falling rapidly but it had only been able to do this by hiring more land and labour (Table 13.1). Even after 1929 the area under cane was expanded. After 1931 the area under cane and the number of workers had to be reduced much more than would have been necessary if the producers had not been tempted by the increased yield per hectare to expand their production. This expansion, however had not been sufficient to keep income from exports at the level of 1927 (Table 13.1, column 5). Even as a percentage of all exports, the value of sugar exports declined after 1928 (column 7).

But perhaps the production of sugar could still be a profitable undertaking for the factory owners? Profits had been high and rising since the beginning of the period under consideration, at least until 1924. However, increased productivity and production in 1928 could not stop the falling trend of profits after 1924. In 1930 profits had turned into losses.[17] Nevertheless the area under cane was still expanded somewhat in 1930 and 1931. In that year, however, a large number of employees were fired, stocks started to pile up, and the value of sugar exports was only half that of 1930 (Table 13.1). In 1932 the first reduction of the area under cane took place and more workers were dismissed, but in fact the first really draconian measures were not taken until 1933.

Conclusions

Neither high productivity of land, nor increasing productivity of labour had been able to save the Java sugar industry. Both production and trade were well organized, and the industry was backed up by an excellent research institution. It has been suggested that the refusal to leave the gold standard made it impossible for Java to compete with producers with a more flexible attitude, but that does not explain why the Netherlands Indies' export performance for other products was so much better than for sugar.[18] It must be concluded that the disappearance of the most important markets for Java sugar, due to prohibitive protection, and the lack of alternative outlets, were the ruin of the industry.

References

1. Figures for total global sugar production, 1840–1940 (cane and beet) in M. Moreno Fraginals, *El ingenio; compléjo económico social cubano del azúcar*, Havana, 1978, (III Vols.), III, 35–40; on Java sugar for the same period:*Changing Economy in Indonesia* (CEI), Vol. I (P. Creutzberg), 1975, 63–76; on the share in world production of other Netherlands Indies' export crops: *Statistisch Jaaroverzicht van Nederlandsch–Indië* (SJO) 1941, 302.

2. On the Java sugar industry before 1800 see K.W. Van Gorkom, 'Historische schets van de suikerindustrie op Java',*Tijdschrift voor Nijverheid en Landbouw van Nederlandsch–Indië*, XXIII (1879); on the fortunes of Java sugar between1775 and 1830 see P. Boomgaard, ' Java's agricultural production 1775–1875', in A. Maddison & G. Prince (eds.), *Economic Development and Social Change in Indonesia, 1820–1940*, Leiden, 1987 (forthcoming).

3. Much has been written about the Cultivation System, although a comprehensive study on this subject is still lacking; some recent titles, with special reference to cane, are R.E. Elson, 'Sugar and peasants; the social impact of the Western sugar industry on the peasantry of the Pasuruan area', unpublished PhD thesis, Monash University (Australia), 1979 ; C. Fasseur, 'Kultuurstelsel en koloniale baten; de Nederlandse exploitatie van Java, 1840–1860', unpublished PhD thesis, Leiden, 1975; R. Van Niel, 'The effect of export cultivations in 19th century Java', *Modern Asian Studies*, XV (1981), 25–58.

4. For details on legislation concerning land for sugar–mills see V.J. Koningsberger, ' De Europese suikerrietcultuur en suikerfabricage', in C.J.J. van Hall & C. van de Koppel (eds), *De Landbouw in de Indische Archipel*, 's– Gravenhage, 1948 (IV Vols.), IIA, 324–7.

5. H.E. Levert, *Inheemsche Arbeid in de Java Suikerindustrie* ,Wageningen, 1934, 98–103; H.Ch.G.J. Van der Mandere, *De Java–Suikerindustrie in Heden en Verleden Gezien in het byzonder in Hare Sociaal–economische Beteekenis*, Amsterdam,1928, 17–8, on the 1884/5 sugar crisis and its aftermath in Java; on sugar prices during the 19th century see N. Deerr, *The History of Sugar*, London, 1949/50, II, 531.

6. Koningsberger, ' De Europese ', 290–234.

7. Tio Poo Tjiang, *De Suikerhandel van Java*. Amsterdam, 1923, 18–37; Van der Mandere, *De Java–*, 92–4.

8. Levert *Inheemsche Arbeid* , 177, 216–17; Tio Poo Tjiang, *De Suikerhandel* , 40–5.

9. Levert *Inheemsche Arbeid* , 218–20 , 273–74; *Economisch Weekblad Voor Nederlandsch–Indië*, extra number, December 1932/Januari 1933, 51–8.

10. A.P.E. Korver, 'Sarekat Islam, 1912–16.', unpublished PhD thesis, Amsterdam,1982, 109–111; A. Van Schaik, 'Colonial control and peasant resources in Java; agricultural involution reconsidered.', unpublished PhD thesis, Amsterdam, 1986, 81–8, 106–14.

11. Levert, *Inheemsche Arbeid* , 114–130.

12. *Ibid.*, 213–15.

13. *Ibid.*, 255–58; Koningsberger, ' De Europese ', 383–84; ; Van der Mandere, *De Java–*, 113.

14. Koningsberger, ' De Europese ', 318, 347–52.

15. J.J. Tichelaar, *De Java–suikerindustrie en Hare Beteekenis voor Land en Volk*, Surabaya, s.a.(1926), 72.

16. Deerr, *History*, II, 587.

17. CEI, Vol. VII (W.L.Korthals Altes), 1987, Tables 4A and 5; *Kroniek van Sternheim*, IX (15.1. 1931), No. 237, 183.

18. The role of the gold standard is discussed in W.L. Korthals Altes, 'The Netherlands Indies and the gold standard during the 1930s', unpublished paper for the

conference on 'The Economics of Africa and Asia During the Inter–war Depression', London, 1985.

14

The Oligopolistic Structure of the Philippine Sugar Industry during the Great Depression

Yoshiko Nagano

The interwar period witnessed a drastic change in the structure of the Philippine sugar industry, both in terms of its export market and productive organization. Before 1898 the Philippines produced muscovado (non–centrifugal) sugar which was shipped mainly to Asian countries such as China, Hong Kong, and Japan. However, after colonization by the United States a reciprocal treaty gave the country free entry into the large US sugar market. This in turn spurred the modernization of the sugar manufacturing sector, first by Americans in the early 1910s and then by local Spaniards and Filipinos during the 1920s. The most striking technological development over these years was the establishment of sugar centrals or mills equipped with the latest equipment, such as vacuum pans, centrifugals, etc.. At the same time, mechanization in sugarcane agriculture remained generally stagnant due to the existence of excess labour in the rural areas.[1] Thus, the productive structure of the industry was made up of haciendas (landed estates) which had been the centre of sugarcane agriculture since the late 19th century and now supplied cane to the newly–erected central mills owned by foreign or local entrepreneurs. Moreover, it is important to note that despite foreign political and economic intervention it was local entrepreneurs and hacendados (landlords) who continued to control the majority of capital as well as land in the sugar industry in the 1920s. This situation was to be reinforced in the subsequent decade. In this chapter, we will discuss the nature of the oligopolistic structure or localization in the ownership of sugar centrals and haciendas, the major characteristic of the Philippine sugar industry during the American colonial period.

The changing export market in the interwar years

Although the United States took possession of the Philippine Islands on signing of the Paris Treaty with Spain in 1898, it was only after the enactment of the Payne–Aldrich Tariff Act of 1909 and the Underwood–Simmons Tariff Act of 1913 that Philippine sugar exports to the United States began to grow substantially. While the Payne–Aldrich Act permitted the free entry of Philippine sugar within the limit of 300,000 tons, the Underwood–Simmons Act removed former restrictions on the entry of Philippine sugar, thereby making it tax–exempt.

As shown in Graph 14.1 the Philippines exported 100,000 to 200,000 metric tons of sugar per year between 1899 and 1913. With the outbreak of WWI, the amount of sugar sold increased to 300,000 tons due to world shortages. However, it was in the 1920s that the full scale expansion of the Philippine industry got underway, with the final shift out of muscovado into centrifugal sugar. Exports rose to 500,000 tons during the early 1920s, 85 per cent of which went to the United States. Subsequently, the total grew to 800,000 tons, with the US share increasing to 90 per cent. The 1920s, therefore, saw centrifugal sugar production firmly established and the country's industry becoming almost entirely dependent on the market of its colonial masters. Besides sugar the Philippines exported manila hemp (abaca), copra, coconut oil and tobacco during the American Period. It is noteworthy that the export value of sugar exceeded that of manila hemp in the 1920s, thereby making it the country's single most important industry.

During the depression of the 1930s, the dependence of Philippine sugar on the US market became much clearer, particularly after the United States introduced a quota system to regulate both the importation and domestic production of sugar.[2] In the 1930s the United States passed two sugar acts, in 1934 and 1937, by which the Philippines was allocated a quota of approximately 1 million short tons each year. However, another regulation was applied under the Tydings–McDuffie Act of 1934 which limited the free entry of Philippine sugar to a maximum of 800,000 long tons of raw sugar and 50,000 of refined sugar. The Philippine producers did not ship more than this tax–exempt limit.[3] But, the Philippines was assured an export market of about 1 million short tons every year during the depressed 1930s, although in 1935 only 500,000 short tons were exported because the quota had been exceeded in the previous year. Despite this privileged market access, in the 1930s the Philippine sugar industry suffered from low world prices and the limitations placed on production. It was also during this period that American capital began to retreat from the country in anticipation of the future independence of the Philippines, a central provision of the Tydings–McDuffie Act.

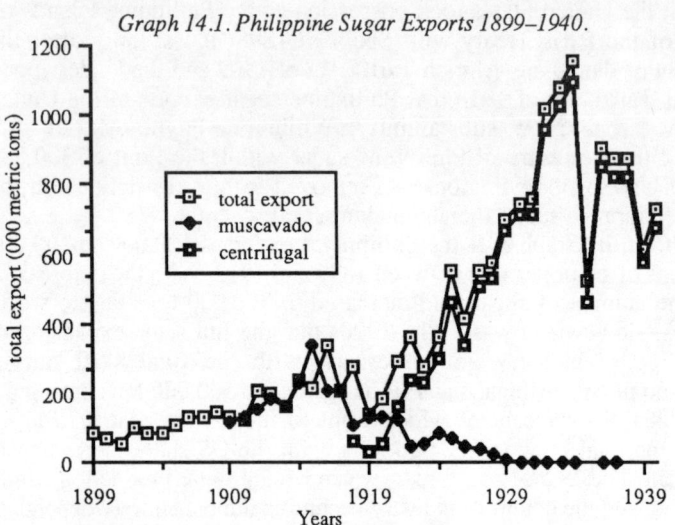

Graph 14.1. *Philippine Sugar Exports 1899–1940.*

Source: Philippine Islands and Philippines (Commonwealth) Bureau of Customs, *Annual Report of the Insular Collector of Customs*, Manila, various years.

Introduction of the quota system in the Philippines

How was the quota system introduced in the Philippines? To impose an export production quota in the country it was necessary for the colonial government to regulate the production of the sugar centrals or sugarcane farms, classifying their production either for export or the domestic market.

Upon the enactment of the 1934 Sugar Act, the Philippine government, following the guidelines of the US Department of Agriculture, set up the Sugar Quota Administration whose duties were to carry out a nationwide survey of sugar centrals and sugarcane farms and to provide a three–part quota. This comprised an amount for the US market, another of domestic use and a final quantity for reserve stocks. These three different quotas were allocated proportionally to the 47 mill districts, each of which was an administrative producing unit composed of one sugar central and from a dozen to several hundred farms producing sugarcane.

In 1939/40 the total quota for the country's producers was 1,107,000 short tons, 86 per cent of which was for export to the US and 14 per cent for the local market and reserve. There was a wide variation among the various districts. For convenience of discussion, we categorize those mill

172

districts whose quota was more than 5 per cent of the total as 'A district', 2 per cent–5 per cent as 'B district', and less than 2 per cent as 'C district'. Of the 47 districts there were 6 A districts, 11 B districts, 28 C districts (The remaining two districts had completely lost their quotas because of shut–down or merger). In total, the A and B districts composed about 80 per cent of the quota, while the remaining 20 per cent was shared by the more numerous C districts.[4] With a single central in each district, these figures clearly indicate the oligopolistic structure of sugar manufacturing in the late 1930s.

Table 14.1 Distribution of Sugarcane Farms by Export Quotas to US.

Export quota to the US			Number of sugarcane farms	
			(piculs)	(%)
0	to	100	15694	65.3
100	to	1000	5933	24.7
1000	to	2000	908	3.8
2000	to	5000	830	3.5
5000	to	10000	397	1.7
10000	to	15000	138	0.6
15000	to	20000	56	0.2
Above 20000			64	0.3
Total			24020	100

Note: The export quota for each farm is calculated assuming that the Philippine export quota is one million short tons.
Source: George H. Bissinger, 'Phillipine Sugar Control IV: Range and Distribution of Production Allowance', *Sugar News*, vol. xviii, no.7, (July 1937), 246.

Did the same level of concentration exist in the agricultural sector? Since a certain percentage of export, domestic and reserve quotas were distributed to every farm based on production an examination of the distribution of the export quota, which was nearly 90 per cent of the total, provides a fairly reasonable idea as to the relative importance of different sized farms. This is shown in Table 14.1. The table shows that the number of farms with a quota of less than 100 *piculs* (1 *picul* = 63.25 kilograms) comprised 65 per cent of the total, those between 100 and 1000 *piculs* 25 per cent, 1000 to 10,000 9 per cent and more than 10,000 only 1 per cent. These figures show the coexistence of a great many small farms with a few very large ones in the 1930s. But to gauge their relative importance vis à vis total output, we must convert these data into production shares. If we assume that the average amount attributed to those farms with quotas in the two ranges above 10,000 *piculs* lay at the midpoint within those ranges and that the 64 farms having over 20,000 *piculs* averaged quotas of only 20,000 (which must give a downward bias

Table 14.2. Nationality of Controlling Interest in Sugar Centrals at the End of the 1930s.

Centrals and controlling interest	Quota (%)[1]	Centrals and controlling interest	Quota (%)
Filipino	**40.67**	**American**	**32.28**
Lizares	*9.88*	*California*	*10.62*
Talisay–Silay Milling Co.	4.54	Pampanga Sugar Mills	6.09
Bacolod–Murcia Milling Co.	4.41	Calamba Sugar Estate	4.53
Central Azucarea del Danao	0.93	*Ossorio*	*10.03*
Montilla	*8.58*	North Negros Co.	5.82
Ma–ao Sugar Central Co	4.23	Victorias Milling Co.	4.21
Isabela Sugar Co.	2.87	*Hawaiian and others*	*11.63*
Ormoc Sugar Co.	0.66	Hawaiian–Philippine Co.	5.54
Rosario Sugar Mills	0.44	San Carlos Milling Co.	3.58
Bataan Sugar Co.	0.38	Bogo–Medellin Milling Co.	0.89
de Leon, Hizon, Lazatin, Rodriguez etc.	*6.78*	Cebu Sugar Co.	0.75
Pampanga Sugar Dev. Co.	6.78	Luzon Sugar Co.	0.49
PNB	*4.68*	Hind Sugar Co.	0.38
Binalbagan Estate	4.68		
Lopez	*3.66*	**Spanish**	**27.04**
Lopez Sugar Mill Co.	1.77	*Elizalde*	*9.94*
Central Santos–Lopez Co.	1.26	Central Azucarera de la Carlota	7.01
Philippine Starch & Sugar Co.	0.63	Central Azucarera de Pilar	1.38

cont.

Ramirez, de Leon, Zobel, Soriano [2]	*2.56*
Central Luzon Milling Co.	2.56
Lizarraga	*2.05*
Kabankalan Sugar Co.	1.09
Mt. Arayat Sugar Co.	0.96
de la Rama	*0.90*
Central de La Rama (Bago)	0.47
Central Leonor	0.43
Central Lourdes	—
Central de la Rama (Talisay)	
Cojuangco	*0.87*
Paniqui Sugar Mills	0.87
Teus	*0.37*
Central Azucarera del Norte	0.37
Cu Unjieng	*0.34*
Mabalacat Sugar Mills	0.34
Buencamino	*0.00*
Nueva–Ecija Sugar Mills	—

Central Sara–Ajuy	0.65
Phillipine Milling Co. (owned by the Archbishop of Manila)	0.90
Tabacalera	*9.76*
Central Azucarera de Tarlac	5.86
Central Azucarera de Bais	3.90
Roxas	*3.36*
Central Azucarera Don Pedro	3.36
Garcia	*1.57*
Asturias Sugar Central	1.57
Vidaurrazaga, Mota	*0.97*
Central San Isidro	0.97
Serra	*0.85*
Central Palma	0.85
Dominicans	*0.46*
Philippine Sugar Estates Dev. Co.	0.46
Ayala	*0.13*
Central Azucarerade de Calatagan	0.13

Notes : [1] the quota for 1939–40. [2] Zobel and Soriano are Spanish. According to some sources the Corbitarte and Ayala families were owners of the mill.

Source : Mainly based on various issues of *Sugar News*, 1919–1941.

to the results) then the proportion controlled by these farms would have been 27.8 per cent. In other words about one per cent of the sugarcane farms took nearly 30 per cent of the quota. This, in turn, is a clear indication of the high degree of productive concentration within this sector of the industry.

Nationality of the controlling interest in centrals

As mentioned above, there were 47 sugar centrals at the end of the 1930s, however, there were wide variations in quotas allowed, and only one–third of the mills can be said to have played a significant role in this sector of the sugar industry. Table 14.2 shows the percentage of the quota of 46 centrals in 1939/40 by specific interest group and nationality.[5] Although in order of relative importance by nationality, the Filipino controlled centrals (41 per cent), were followed by the American (32 per cent) and the Spanish (27 per cent), in terms of average size it was the US owned mills which came first with 3.2 per cent of the quota followed by the Spanish (2.3 per cent) and finally the Filipino centrals (1.7 per cent). However, as will be shown below, within each group there was a further important difference in relative size.

Filipino centrals

Among the 24 centrals controlled by Filipino capital, five held 4 to 6 per cent of the total quota, two were assigned 2 to 3 per cent and the remaining seventeen had less than 2 per cent each. The degree of concentration here is shown by the fact that 61 per cent of the quota for Filipino owned mills was in the hands of only five centrals, which together had nearly 25 per cent of the entire industry's quota.

Together with the Isabela Sugar Co., these five centrals were known as the 'Bank Centrals'. They were initially incorporated in the late 1910s or early 1920s, by prominent landlords or planters on the island of Negros or Central Luzon with the financial assistance of the Philippine National Bank (PNB). They were given long–term loans from the PNB, however, all six centrals were soon in financial trouble and found themselves taken over by the PNB in the early 1920s. It was only at the end of the decade or in the early 1930s that ownership or management was returned to the private sector,[6] which in this case meant just a few prominent families. In fact, in the 1930s both the Lizares and the Montilla families became most powerful Filipino sugar producers. By late in the decade the former family controlled three mills, while the Montillas added a further three mills to their empire in the mid–1930s, which by 1940 comprised five mills. Thus, concentration in the ownership of sugar centrals accelerated during

176

the depression, smaller mills being taken over or ceasing operation (Table 14.2).

American centrals

Among the 10 centrals under the control of US capital, the six largest had 90 per cent of the American centrals' quota, or 30 per cent of the entire country's. These six were in turn owned by three groups which were directly or indirectly linked to US sugar interests.

The first group, based in California, controlled the Pampanga Sugar Mills and the Calamba Sugar Estate. The latter was organized in 1912 by a consortium headed by Alfred Ehrman, who obtained 7200 hectares in the province of Leguna from the Philippine government.[7] In 1919, this estate,together with the Spreckels family, set–up the Pampanga Sugar Mills, in the western part of Pampanga province.[8] These California interests probably saw investment in sugar manufacturing to be profitable because of the preferential tariff available for the US market. They also acted as sugar brokers in selling their own sugar in this market. However, in 1941, foreseeing the dark future for the industry, the Calamba Sugar Estate was sold to the Madrigal family, Filipinos prominent in shipping.

The North Negros Sugar Co. and the Victorias Milling Co. Inc. were incorporated both during and after WWI by Miguel J. Ossorio, a Spanish Filipino. However, when Ossorio acquired US citizenship in the late 1920s, the nationality of the capital represented by the two mills also changed.[9] It is noteworthy that in the late 1930s the North Negros Sugar Co. was one of the stockholders of the American Sugar Refining Co. and the Great Western Co., both large US companies. This strongly suggests that Ossorio's centrals probably had close and intimate links with US sugar interests.[10]

The other two mills were controlled by Hawaiian capital. The San Carlos Milling Co. Ltd. was the first Hawaiian central and was established in Negros in 1912. It was first managed by the Welch–Fairchild Co. and later in the 1930s by the Bishop Trust Co.[11] The former company also managed the other Hawaiian central, the Hawaiian–Philippine Co. created in 1918 in Negros under the initiative of the Hawaiian Sugar Planters' Association (HSPA). The HSPA as well as individual–based capitalists had been interested in investing in the Philippine sugar industry, the latter as part of its efforts at direct recruitment of Filipino labour for the Hawaiian sugar industry. The Hawaiian–Philippine Sugar Co, was promoted by the HSPA in response to the pressure from the Philippines to set up a modern mill and not use its properties simply as a source of labour.[12]

Spanish centrals

Of the 12 Spanish centrals, only four had more than 3 per cent of the industry's total allocation. They were controlled by two families and one large corporation. The Elizalde family was prominent in commerce in Manila and had owned haciendas since the late 19th century.[13] In the late 1930s, it owned three mills and managed a fourth under contract with the Archbishop of Manila. The fourth mill, the Philippine Milling Co., was formerly the Mindoro Sugar Co., established originally by such famous American sugar interests as H.O. Havemeyer, C.J. Welch and C.H. Seneff. It was the first sugar central in the Philippines and was established on the 22,500 hectare San José Estate. Operations began in 1912/13. However, being located in Mindoro where labour was scarce, it did not produce enough sugar to recover its initial investment, and thus, it changed hands and was finally sold to the Archbishop of Manila in 1929.[14]

The Roxas family owned both the Central Azucarera Don Pedro and the Central Azucarera de Calatagan. Both were in the province of Batangas, where they had had big sugar haciendas since the 19th century. However, in 1930 the latter was taken over by their relations in the Ayala family under the terms of a bequest.[15] Finally, the Tabacalera (Compañía General de Tabacos de Filipinas), the largest tobacco company in the country, established the Central Azucarera de Bais in Negros in 1918 and its other mill in Central Luzon in 1927. The latter was set up within the company's own 10,000 hectare Hacienda Luisita,[16] which was obtained by the Tabacalera in the early 1880s. Its original object had been to plant tobacco in the hacienda, but it failed. In the 1910s it started to grow rice and with the establishment of the mill 5,700 hectares were devoted to planting sugarcane by the mid–1930s.

Rivalry between millers and planters

One of the principle characteristics of the Philippine sugar industry is the large number of independent sugarcane farms and few central mills. Although there were some exceptions, and while some planters also owned mills, the interests of millers and planters remained essentially distinct. The latter were mainly Filipinos of Spanish or Chinese mestizo origin together with a number of Spaniards. With few exceptions, Americans were not interested in owning land. This was because the Public Land Act of 1903 limited ownership to 16 hectares for individuals and 1024 hectares for corporations. Furthermore, much of the sugar producing land was under Filipino control and had been since the late 19th century. In the two cases where US interests did own sizable holdings, the Calamba Sugar Estate and the San Jose Estate, this was because the land in these cases had been under the control of the church and was, therefore, not considered as public lands under the 1903 act.

To obtain cane most of the newly established centrals had to come to terms with planters who produced muscovado sugar in their own small mills. They, therefore, had to offer a more profitable alternative. At first, a simple cash payment was tried, but this was not successful. A share system, similar to one used in Cuba and first introduced by the San Carlos Milling Co. Ltd. in Negros proved more popular. Under this procedure the sugar produced was divided in agreed upon proportions between millers and planters.[17] The planters directly benefited from the much more efficient milling and manufacturing practices of the larger more modern mills and also did better because a higher, export quality of sugar was produced. The share system thus spread to most of the islands' sugar producing regions and by the 1920s had become predominant.

Although it might be thought that under such a system the interests of millers and planters would be harmonized, this was not the case. This was particularly true during the hard years of the 1930s. Because of the extremely low sugar prices many small planters became heavily indebted to centrals or the banks. In response the government set up the National Sugar Board whose main function was to revise the percentage of sugar shared between the two groups. The aim was to try to improve the planters' position, as they were at a substantial disadvantage vis–à–vis the centrals who were generally the sole cane buyers within each sugar region.

According to a nation–wide survey conducted by the Board in the 1930s the percentage of sugar taken by the millers in Luzon was 50–65 per cent and in Negros from 39–45 per cent. The rate of profit varied much more dramatically. In Luzon planters made between 7 and 15 per cent, while millers earned from 3 to 50 per cent. In Negros the percentages were respectively 10 to 22 per cent and 6 to 40 per cent. Because of this clear inequality in return, the Board proposed a new scheme which would have reduced the millers' share to between 30 to 54 per cent in Luzon and 31 to 52 per cent in Negros.[18] However, this was not to be implemented until after the war.

The battle between millers and planters was only one aspect of the rivalries within the sugar industry. The other was the grievances and poverty of tenants and hacienda workers who were cruelly exploited through high rents and low wages. This is a very complex story which can only be briefly mentioned here. It was in 1938–39 that the peasants and workers in Pampanga Province of Central Luzon and in Iloilo Province of Western Viscayas began to demand higher wages and fairer treatment. Particularly in Pampanga, the strikes of peasants and workers in the sugar–producing regions soon became integrated into a larger mass struggle which was to spread to most of the provinces of Central Luzon.

Conclusion

In this extremely brief survey of conditions in the Philippine sugar industry during the Great Depression a number of salient points emerge. It is clear that in the sugar–producing regions landholding was already concentrated during the era of Spanish control. After US colonization some Americans bought plantations and occupied large tracts of land. However, most of the cane lands remained under the control of Filipinos. Because of this and other reasons the newly–established sugar centrals had to introduce a share system to obtain the cane they required.

The main feature of the sugar industry during the interwar period was the establishment of sugar centrals. This was a trend which developed in tandem with the opening of the US market to Philippine sugar. Since refineries in the United States wanted centrifugal not muscovado sugar it was imperative for modern mills to be established. However, the Philippines was remote from the United States, and there were several factors which discouraged investment. These included the Jones Law of 1916 which promised future independence and the 1903 restrictions on landowning. Because of this it was only a handful of US entrepreneurs, and a greater number of Filipinos and local Spaniards who invested actively in the milling sector under government initiative.

It was during the 1930s that the trend for the concentration of mills was accelerated. Because of low prices, by the end of the decade many of the smaller mills had either been bought out by larger ones or simply ceased operation. After WWII many more small mills were to disappear. At the same time, local ownership of the mills became more pronounced. This was due in part to the introduction of the quota system in 1934 and the assurance of independence after the Commonwealth period. It may seem contradictory, but the introduction by the US of the quota system worked directly against US investment in the Philippine sugar industry and instead encouraged local entrepreneurs and planters. This was because the trade policy regarding sugar was formulated with US domestic producers and market conditions here as the primary considerations. Whatever the causes, the result was to strengthen local sugar interests which soon developed into a powerful economic and political oligarchy within the Philippines. This in turn set the stage for the further consolidation of their power in the 1950s.

References

1. See above, chapters by Albert and Beechert.

2. See above, 'Introduction'.

3. See Graph 14.1. 800,000 long tons of raw and 50,000 long tons of refined are equal to about 970,000 short tons.

4. '1939–40 Allotments', *Sugar News*, XX, 9, Sept. 1939, 404.

5. The missing mill, operated by the University of the Philippines' College of Agriculture in the Laguna province of Southern Luzon is excluded, since it was mainly used for experiment and its output was negligible.

6. *Annual Report of the Philippine Sugar Centrals Agency*, nos. 1–5, 1922–26; *Handbook of the Philippine Sugar Industry*, 2nd ed., Manila, 1929, 70, 91, 135, 150, 165, 187.

7. *The National Cyclopedia of American Biography*, New York, 1945, XXXII, 145.

8. John A. Larkin, *The Pampangans: Colonial Society in a Philippine Province*, Berkeley, 1972, 284–85.

9. Kunio Yoshihara, *Philippine Industrialization: Foreign and Domestic Capital*, Quezon City and Singapore, 1985, 66–7.

10. Office of the President, National Sugar Board, ' Memorandum on the Sugar Industry...', (unpublished), Manila, 1939, 121.

11. *First Report of the San Carlos Milling Company Limited*, Honolulu, 1914, U.S. National Archives, BIA 25996 7–A.

12. *First Report of the Hawaiian–Philippine Co. Fiscal Year Ended Sept. 30, 1921 and Sept. 30, 1922*, 14–15.

13. Carlos Quirino, *History of the Phillippine Sugar Industry*, Manila, 1974, 69.

14. Macario Z. Landicho, *The Mindoro Yearbook: 1901–1951*, Manila, 1952, 235.

15. 'Calatagan Mill Bought by Thai Concern', *Sugar News*, XXII, 11, Nov. 1941, 447.

16. Emili Girat Raventos, *La Compañía General de Tabacos de Filipinas, 1881–1981*, Barcelona, 1981, 148–49.

17. Yves Henry, *Conditions techniques et financières de la production de sucre aux Philippines*, Hanoi, 1928, 129.

18. Taiwan Ginko, Taihoku, *Hito Togyo Chosa Hokoku (Hito Togyo Shinsakai)*, 1944, 14–18 [Taiwan Bank, Taipei Research Section, *The Report of the Survey on the Philippine Sugar Industry (National Sugar Board)*].

15

Sugar Mills and Peasants in Northern India, 1914–40

Shahid Amin

In the extant literature on sugar the 'Indies' are usually taken to mean the British West or the Dutch East Indies, primarily because there was not much of a sucrose content to the brightest jewel in the British crown. India was by no means a sugar colony. It did not have large factory estates or capitalist farms, central mills did not emerge as an important consumer of sugarcane till quite late in the 1930s, peasant produced raw sugar (*gur*) rather than vacuum pan white sugar remained the predominant commodity form, the product did not cater for the world market, and most significantly, the culture of sugarcane was characterized by small peasant production.

Yet India demands attention. Compared to other colonies the area under sugarcane in the 1930s was a sizeable 3.4 million acres, most of it concentrated in the northern state of U.P. (Uttar Pradesh), and though cane acreage in India as a whole represented a meagre 1.8 per cent of the total cultivated area, the culture of the crop was of crucial significance in regions of its specialized cultivation. The interwar years saw a dramatic transformation of the Indian sugar industry, as a rapid proliferation of vacuum pan mills, erected behind tariff walls in the middle of the Depression years, took place. However this did not lead to a supplanting of petit culture, nor did it bring about any marked change in the condition of the majority of the producers. The structure that had sustained and borne heavily upon small peasant production lay undisturbed by the change of context.

The burden of this chapter is to relate the history of white sugar production in the interwar years to the structure of production in the villages of U.P. in general and the district of Gorakhpur in particular.

Table 15.1. Import of White Sugar and Production of Sweet–Stuffs in India 1908/09 to 1929–30 (000 tons).

	+Gur	Factory Cane Sugar	Imports	Total	% of imports to total
1908/09 to 1910–11	2,161	Negligible	562	2,723	20.6
1911/12 to 1913–14	2,435	"	625	3,060	20.4
1914/15 to 1916–17	2,585	15*	436	3,036	14.3
1917/18 to 1919–20	2,864	20	411	3,295	12.2
1920/21 to 1922–23	2,671	26	409	3,106	13.2
1923/24 to 1925–26	2,893	42	582	3,517	16.6
1926/27 to 1928–29	2,996	68	793	3,827	20.7
1928–29	2,634	68	859	3,561	24.1
1929–30	2,681	68	1,003	3,752	26.7

Note: + Raw Sugar; * Estimated.
Source: Indian Tariff Board: *Report on the Sugar Industry*, Evidence 1, 19–21; 39–40 (Calcutta, 1931–32).

I take Gorakhpur as my core area because with over 200,000 acres under cane and two–dozen mills to account for it, the district represented the domination of capitalist millers over peasant producers in an accentuated form.

The prewar history of the Indian industry is one of slow development of a few beleaguered European enterprises, followed by rapid growth of Indian owned central mills, occasioned principally by the grant of tariff protection in 1932.[1] From the late nineteenth century Java was the main exporter of white sugar to India. Even after incurring freight charges and an import duty of 6 annas a *maund* (82.5 pounds). Java could undersell the small amount of factory sugar that was being produced by English and Scottish business houses in central and eastern U.P. and the neighbouring districts of north Bihar. This was the position on the eve of WWI. The war effectively protected the nascent industry against foreign competition and gave a stimulus to white sugar production. After the war with an over–production of world sugar and the erection of tariff barriers in the chief importing countries India remained one of the few 'free areas' where Javanese sugar found a lucrative and substantial market. During the 1920s imports were one again up, accounting for a quarter of the total sweetstuffs available in India. Import duties (with revenue considerations in mind) were repeatedly raised from 15 per cent *ad valorem* in 1916 to 25 per cent in 1922 and in effect to 50 per cent in 1925. But, despite this between 1921/2 and 1929/30 the price of Java sugar in the Calcutta market was very nearly halved. In 1929/30 a record 1 million tons of sugar was imported into India, over 70 per cent of this came from Java. During the late 1920s India took about 40 per cent of Java's total exports.[2]

Until the grant of tariff protection in 1932 the colonial state, apart from raising revenue duties, did nothing very much to aid the growth of central mills in India. To be sure, in the wake of the war a Sugar Committee had been appointed. It toured Java and the Indian provinces, bemoaned the smallness and inefficiency of Indian factories and of peasant cane farms, without coming up with any radical solutions. It ruled out acquisition of peasant cultivated lands for central mills, and saw no pressing need to upset the prevailing *laissez faire* orthodoxy by recommending tariff protection. A sugarcane breeding centre had been set up earlier in 1912, however in the absence of a powerful mill sector the newer strains spread quite slowly before the mid 1930s.

In 1931 the colonial government afforded tariff protection to the sugar industry for a period of fifteen years. The effective rate of duty was around 185 per cent *ad valorem* on the basis of Rs.15 per hundredweight of Java sugar in Bombay, including the duty and landing and handling charges. This far reaching step was taken with the problems of liquidity in the villages, shrinkage of state revenues and mounting agrarian upheaval firmly in mind. Sugarcane after all was an important rent paying crop. The creation of a protected market for Indian sugar, it was thought, would give a much needed relief to an ailing countryside hit by the whirlwind of the world depression.[3] During the Great Depression relative agricultural prices moved in favour of sugarcane. It was not that prices paid by the factories for cane did not fall, in fact the drop in Gorakhpur district in 1931/2 over 1928/9 was of the order of 38 per cent, it was just that this fall was 12 to 20 per cent lower than that for the superior and inferior grains and also less than that for raw sugar.[4] The sudden expansion in cane acreage in Gorakhpur for mills was accomplished with relative ease. There were no technical barriers to an expansion of cane cultivation. Irrigation was not a constraint in the water rententive soil of the eastern part of the district, nor did peasant households have to bother about converting the increased output of cane into *gur* (raw sugar). Earlier the area devoted to cane was limited by the availability of stock to crush and boil it into *gur*. In the 1930s sugar cane rather than *gur* became the predominant commodity form.

The impact of the tariff protection was phenomenal. Net imports of foreign sugar as a percentage of Indian–made sugar declined from a high 230 per cent in 1930/1 to an insignificant 1.8 per cent in 1936/7. Both acreage under cane and vacuum pan factories registered impressive growth, but these developments were much more in evidence in the province of U.P. and the eastern district of Gorakhpur within it (Table 15.2). Sugar mills in U.P. increased from 15 in 1930 to 75 in 1937, out of these 23 were in the district of Gorakhpur alone. Cane acreage in the district doubled between 1930 and 1936 (from 138,000 acres to 276,000 acres).

Approximately 70 per cent of the marketable crop was now earmarked for the local mills. The rapid growth of sugar mills in India accounted in part for the declining fortunes of the Javanese industry. The irony is that many a sugar factory raising its head for the first time by the way–side railway stations in the interior of Gorakhpur had been bought at giveaway prices in Java.

Central mills, especially in the pre–1937 period, were small and relatively inefficient. The combined production of the 18 factories in operation in 1919/20 was just about equal to the output of three standard Java mills. The problem in India was not simply of milling inefficiency, but also of small total crush per season, which was related both to the duration of the season and the installed capacity of the mills. The original installed capacity in India was rather low. Before 1937 (the year of self–sufficiency in sugar production) the majority of Indian mills operated plants of under 400 tons a day crushing capacity. In Gorakhpur 17 out of the 23 mills in operation in 1934/5 fell within this low limit.

On the eve of state protection, the 'white mill sector', in line with the general trend of post–war industrialization, was already losing ground to the 'native capitalists'. In 1931/2, the European dominated ISPA (Indian Sugar Producers' Association) accounted for just one-third of the total capacity of central mills in the country. The bulk of the white sugar production was now in the hands of Indian capitalists, organized into the Calcutta based Indian Sugar Mills Association (ISMA), a body which remained under the control of the U.P. and Bihar factories until well into the late 1940s. Unlike the situation in the southern province of Madras, the U.P. mills were typically joint stock companies. Ownership of sugar factories was in the hands of a tribe of up and coming Marwari and Punjabi mercantile communities who in the post WWI war period were making a significant entry into the jute, cement and sugar industries. In Madras, merchants and rich peasants, curtailing their grain financing and trading operations during the Depression, attempted to transfer their capital into central mills. However most such firms were family based enterprises and operated uneconomic plants even by the low Indian standards. By the late 1930s the combined daily crush of these pigmy mills was a pathetic 272 tones of canes per day. No wonder these small–time family concerns soon folded. Even in western India, the land where there were to be co–operative mills, it was the Bombay industrialists rather than local rich peasants who were the main entrepreneurs in the immediate post–protection period, although one successful rich peasant, single–caste, mill (the Saswad Mali Sugar Factory) was established in 1934.[5]

In eastern U.P. sugar capitalists failed to acquire large blocks of cane lands either through private leasing or through state support. The Gorakhpur capitalists were particularly reliant on peasant cane production

Table 15.2: *Sugar Mills and Sugar Production, India and U.P., 1930–40.*

Year	1	2	3	4	5	6	7	8	9
1930–31	–	2905	29	10	120	36	200	350	808
1931–32	1593	3077	32	17	159	62	250	471	351
1932–33	1793	3425	57	27	290	78	275	643	384
1933–34	1734	3422	112	16	454	65	200	719	320
1934–35	1840	3602	130	13	578	43	150	771	311
1935–36	2249	4154	135	13	932	48	125	1105	132
1936–37	2469	4582	137	9	1129	25	100	1254	22
1937–38	2222	3997	136	–	940	17	125	1072	–
1938–39	1635	3270	139	–	650	14	100	765	–
1939–40	2062	3788	145	–	1241	31	100	1373	–

Note: Cols 1 and 2: Acreage under Sugarcane (000 Acres), (Col 1, U.P. and Col 2, All-India); Col 3: No. of vacuum–pan factories (All India); Col 4: No. of Sugar Refineries (All India); Col 5: Vacuum Pan Factory Production; Col 6: Gur Refinery Production; Col 7: Khandsari and Open Plan Process   [Cols 5-8 Production of Sugar; Col 8: Total Sugar Production, all-India (000 tons)]; Col 9: Net Imports of Sugar into India (000 tons);
Source: I.T.B: *Report on the Sugar Industry* (1938), Table vi and I.S.M.A:*Report of the Committee, 1939–40*, App. I, Table 4.

and the existing agrarian structure. It was therefore essential for the central mills in Gorakhpur to work out some kind of an arrangement by which they got the right quantities of cane, and at the right time, from a multitude of petty suppliers.

As the mills in Gorakhpur were evenly spread throughout the main cane growing tracts they did not have to haul cane from distant areas. The bulk of the cane supplies were brought from a radius of 10–15 miles from the factory in bullock carts by the peasants themselves. That there should have been some competition among the various factories seems natural enough, for sugar mills in the district were situated in close proximity of each other, and often competed for the same cane from adjoining villages. What is significant is that it was this same competition that seems to have forged the alliance between sugar capitalists and landlords and moneylenders to the detriment of the peasantry. The Indian Sugar Mills Association (ISMA) through its Gorakhpur branch tried zonal and boundary arrangements between the different factories, but this was not always successful. A more general expedient to prevent competition and secure supplies was to bid for cane contractors rather than for the peasants' sugar cane. A three-tier middlemen structure seems to have been in existence in Gorakhpur. An average factory would engage some 60 agents to supply cane to it. In some cases advances would be made to them to be passed on to the peasants (something about which the factories were not unduly bothered). Out of these 60 contractors 8 or so would be very big men, usually *zamindars* (landlords) belonging to the higher castes, who would among themselves account for some 40 per cent of the total cane supplies (each contracting for between 100,000 to 200,000 *maunds* of cane). There would also be a substantial number of middle ranking contractors, also belonging to the higher castes, each accounting for 20 to 75,000 *maunds* of cane. Then came the dozen or so smaller contractors, mainly rich peasants with a surplus of carts, who contracted for 8,000 to 20,000 *maunds*.

The bigger contractors were no creatures of the mills, 'they were', to quote a district officer, '*zamindars* and other persons of local influence who for a consideration ... of 15 per cent of the price of cane undertook to hound their cultivators to a particular factory'. Two things are very clear about these gentlemen. First, that most of the big contractors had strong *zamindari* connections. Second, the factories gave them a commission on the cane supplied through them not for any 'services rendered', but in recognition of their hold over peasant produced cane, the supply of which they could withhold from a particular factory, if they so desired.

In Rohilkhand, (central U.P.) by contrast, the entrenched position of the *khandsari* and the involvement of *zamindars* and moneylender-manufacturers in *khand* (indigenously refined sugar) production made the working of the contractor system a difficult proposition. Landlords and moneylenders here actively tried to prevent the intrusion of central mills into the local village economy which they dominated not simply as rentiers and usurers but also as sugar manufacturers. Generally speaking, the operation of sugar mills in this region was not possible in alliance with dominant rural groups, but only by the capitalists following the 'proto–industrial' path chartered by the indigenous sugar manufacturers of Rohilkhand. Cane supply strategies bore a striking family resemblance to the existing ones perfected by the *khandsaris* at least since the 1850s.

As far as the Gorakhpur peasantry was concerned, the domination of the landlords as rentiers, moneylenders and usurers found a new type of instrument in the *sattas* (coupons) and *purzis* (receipts) through which the mills now procured sugarcane. The factories sought to ensure a regular and steady supply by enforcing a system of coupons without which a peasant would be hard put to sell his cane at the mill gate. However, as these coupons were distributed through the cane contractors, it meant that the peasantry's access to the sugarcane market was still being dominated by landlords and moneylenders. It was not simply a case of *zamindars* selling their tenants, cane as their own. The contractors were also responsible for making the final payment to all the peasants who supplied cane through them. Thus the *purzi*, or cane receipt, became an instrument for the extraction of surplus. Most of the contractors had their 'pay offices' in their villages, where to quote a local officer, 'they make deductions to which they are not entitled'. In fact the same officer went on to define cane contractors as '*zamindars* who confuse rental dues with the price of cane'. The coming of the mills then did nothing to alter the structure of dependence in the villages of Gorakhpur. Instead of the hypothecation of the *gur* produce there was now a pledging of the price of cane.[6]

If the conditions of agricultural production had been responsible for the phenomenon of the cane contractor, the transport of cane gave a fillip to a different kind of an intermediary – the 'professional carter'. Where the peasant himself took his cane to the factory, more often than not it was a hired cart that he was driving. Along with an expansion of cane acreage in the 1930s there was a phenomenal increase in the supply of bullock carts. However this increase was not uniform over all sections of the peasantry. It was basically confined to bigger and richer peasants who, having earlier played a part in the transport of agricultural produce, now invested in additional carts to meet the new requirements caused by the large scale operations of central mills. Transport hire accounted on an average for 15–20 per cent of the price sugarcane, and there might well have been a relative increase in freight for a bulky commodity like cane compared to raw sugar. Just as the mills did not create any new marketing and credit networks but relied on landed and moneyed intermediaries, similarly they seem to have strengthened the position of the 'professional carters' over the small cane–growing peasants. These intermediaries derived their strength from the traditional position of political and economic domination of rural society rather than from their special connection with the factories. Capitalist domination of the cane–growing peasantry was thus contingent on the latter's prior subordination to the landlords, moneylenders and the richer peasants.[7]

So far as the central mills were concerned, there was a straightforward clash between them and the peasants over the pricing of the raw material. The price of cane was roughly half the total cost of sugar, depending on the milling efficiency of individual factories. The object of the sugar capitalists was to reduce the cost of the raw material. This could be achieved to the mutual benefit of both parties only if there were a substantial increase in yields per acre, so that the total net receipts of the peasantry even at reduced prices exceeded the returns from lower yields at a higher price. Improved varieties of cane did raise yields somewhat (even here the least advance was made in Gorakhpur), but net returns to the peasantry were still determined by high rates of interest and consumption loans taken during the lean periods of the harvest cycle.

Apart from the issue of the price of cane, which was of course crucial, there were other major contradictions between the dictates of the peasant's harvest cycle and the preference of the factories for an extended crushing season so as to minimize overhead costs. The peasants wanted to harvest their cane as early as possible and get on with other agricultural operations, while the mills wanted to extend the crushing season. The average number of working days of a sugar plant in India was around 100 in 1930 and as low as 60 in some cases. This compares to the crushing season in Java which spanned an average 126 days. This was because the indigenous canes of north India would not mature before the middle of December, and even at this time they were not fully ripe. Consequently, 'the factories had to crush unripe cane at the beginning of the season, over–ripe canes at the end of the season and normal canes during the middle of the season.'[8] The introduction of early– and late–ripening 'Coimbatore varieties' resulted in an economic staggering of the crushing season. This very advance brought the conflicting interests of the millers and the peasants to the forefront. The capitalists contended that the recovery of sugar from the early crop was very low, making crushing operations uneconomic as recovery reached its peak from mid–March to mid–April. However there was a traditional objection in eastern U.P. to cane standing in the fields beyond the end of March, as the uncut crop (and the attendant problem of transporting it to the mill) interfered with other contemporaneous agricultural tasks. March was the time for sowing the next year's cane crop and harvesting the spring grains.

Another problem with the improved strains was that in the case of early ripening varieties a high sucrose content was generally associated with low tonnage, and in the case of late ripening varieties tonnage declined with the advent of hot season and the consequent drying of cane. In north India, as a rule, the price paid by the mills was based on the weight of cane without taking into account its sugar content, though a system of deductions and premia was worked out on paper in the late 1930s. The mills wanted a greater recovery of sugar, without having to incur the cost of testing the quality of the juice for each and every bullock cart full of cane that creaked up to their weigh–bridge. There was widespread under–weighing of peasants' cane even after the appointment of cane inspectors in 1933 to keep a check on such malpractices. In fact the Chief Cane Inspector of Gorakhpur was of the opinion that the high sugar rates shown by some mills were simply the result of their crushing more cane than they recorded as having paid for, i.e. by 'persistent swindling' of the cane peasants.

The small peasant was not interested in the percentage of sugar in his cane. For him the incentives for taking to improved varieties were three: greater yield per acre, early ripening (of a long–standing crop) which would help him get rid of the cane quickly, and relative immunity from disease and pests. The factories also wanted an increase in yields. But above all they desired an increase in the sucrose content, and the dissemination of early, middling and late ripening varieties among the peasantry. These two requirements were incidental to the peasants' calculations, for now that their weak bullocks did not have to crush the cane for the manufacture of *gur*, they wanted a weighty as well as an early ripening cane. Here was a clash of interests based on the contradiction between the role of sugarcane in peasant agriculture and its industrial use by the mills. Given the role of sugarcane in the annual crop cycle of the peasantry, there were limits to the peasants growing cane varieties which interfered with this rhythm.

As for the rate of payment for cane, this was not left entirely to the forces of supply and demand. A scale of prices for sugarcane was evolved to ensure that the capitalists shared the benefits of the protection with the peasantry. A formula was worked out in 1934 by which cane prices were periodically related to the ruling price of white sugar in the central market of Kanpur. This was the minimum price schedule, the floor below which cane prices should not fall. The sliding scale of 1934 was not entirely fair, for the relative share of the peasantry decreased with every increase in the price of sugar. This flaw was rectified by the nationalist Congress government when it was briefly in office in U.P. and Bihar in the late 1930s. However, though the nationalists, from the provincial leadership down to the local Congress leaders–turned–legislators, sought to mediate between the capitalists and the peasants in parliamentary and extra–

parliamentary ways, their interventions were not sufficient to change the structural basis of dependent peasant production in 'mill areas' like Gorakhpur. Without going into the details of this politicization of cane supply, as the capitalists called it, let us just concentrate on cane–price formation in two very different sugarcane regions of north India.

In areas like Meerut, in western U.P., where the peasantry was in a strong bargaining position, helped for instance by the alternative outlet of superior quality edible *gur*, the minimum price schedule was largely irrelevant. Here the price paid by the mills was higher than the prescribed minimum because they had to induce the peasants away from *gur* manufacture at the going rate. In Gorakhpur, by contrast (a typical mill area for which the price schedule was ideally meant), the actual price paid hovered round the minimum permissible which was intended to be the floor rather than the ceiling. The absence of a bargaining counter in *gur* production (which had withered away under the impact of central mills) and the almost total domination of cane cultivation by the local factories, meant that the Gorakhpur peasantry lost most during years of over–production (when surplus cane would have to be sold at a loss or even burnt to make room for the next crop), and correspondingly did not gain all that much during years of underproduction.

Co–operative marketing of cane, especially in mill–dominated areas, was considered a major bulwark of the minimum price legislation, one that would ensure that the peasants actually received what the government decreed they should. When the Collector of Gorakhpur queried the effectiveness of the new schedule in his district in 1934, he was informed, '... If the growers know what the fair price is they can organize themselves in co–operative societies for the supply of cane at the standard price and this in turn will mean that the price will be enforced.' However this was easier said than done. Even under the U.P. and Bihar Sugar Factories Control Acts (1938), by which the Congress governments of these two adjoining provinces ensured a system of regulated cane supplies in specifically demarcated zones to individual factories, the weak bargaining position of the smaller peasants in the villages and at the factory gates was far from completely overcome. A major effort was made under the new dispensation to limit the freedom of the capitalists and the cane contractors. The predatory practices of the factories in matters of weighing were curtailed to a great extent. Cane contractors were outlawed from 'reserved areas'. The official support of co–operative marketing and the creation of the supply unions as legal persons ensured that the safeguards had now a greater chance of success. What a small peripatetic squad of Cane Inspectors operating from the district headquarters did not earlier achieve could perhaps be accomplished by the two dozen Co–operative Unions of Gorakhpur.[9]

However, it is worth reiterating that the subjection of the small peasantry to the mills and the erstwhile contractors was not a result of marketing relations alone. It also reflected their dependence on the dominant social classes in the villages and on the yearly production cycle. In as much as the new arrangements did not fundamentally alter this structure of dependence at the production level, but merely sought to reform the marketing structure above it, there were bound to be instances of these 'safeguards' failing to be effective in practice. Coupons were issued by the staff of the societies which surveyed the cane area, specifying the quantities which individual peasants were supposed to sell through the co–operative on particular days to a specified factory. A special report in 1940 maintained that 'the more influential zamindars (landlords) and moneylenders often manage to win over surveyors, and through the extra purzis (coupons) received...sold the cane of their weaker tenants as their own in order to realize their rent and other dues'.[10] Given the importance of sugarcane in the appropriation of peasant surplus, it is hardly surprising to find that the internal functioning of these co–operatives also articulated the structure of agrarian society. It was and is the paradox of rural India that dealings between capitalist buyers and peasant sellers can be direct and above–board only through the creation of huge bureaucratic machineries, which are in turn taken over by the dominant sections of the rural population.

References

1. Amiya Bagchi, *Private Investment in India, 1900-1939*, Cambridge, 1972.

2. See Table 15.1. See also Indian Tariff Board, *Report on the Sugar Industry* , Calcutta, 1931, 22-3; Prinsen Geerligs, *Cane Sugar Production, 1912-37* , London, 1938, 83-4.

3. *Report on Sugar Industry*, 1931, 40.

4. Revenue Dept. File 29/1930-1 and 14/1931-2, Commissioner's Records, Gorakhpur. See also R.K. Bhan, 'The recent economic depression in India, with special reference to agricultural and rural life', unpublished Ph.D. thesis, London School of Economics, 1940.

5. Bagchi, *Private Investment*; ; Sanjaya Baru, 'State and Industrialization: Political Economy of Sugar Policy, 1932-47', *Economic and Political Weekly*, xviii: 5 (29 Jan. 1983); Christopher Baker, *An Indian Rural Economy, 1880-1955: the Tamilnad Countryside* , Delhi, 1984, 375;

Donald W. Attwood, 'Peasants versus Capitalists in the Indian Sugar industry: the Impact of the Irrigation Frontier', *Journal of Asian Studies,* xlv: 1 (Nov. 1985).

6. Shahid Amin, 'Peasants and Capitalists in Northern India: Kisans in the Cane Commodity Circuit in Gorakhpur in the 1930s', *Journal of Peasant Studies*, viii: 3 (April 1981).

7. See Shahid Amin, *Sugarcane and Sugar in Gorakhpur: An Inquiry into Peasant Production for Capitalist Enterprise in Colonial India*, Delhi, 1984, where the arguments presented in this and the following paragraphs are developed in greater detail.

8. *Report on Sugar Industry*, 1930, ii, 537.

9. For details, see Amin, *Sugarcane in Gorakhpur*, chapters 8 and 9.

10. *Report of Sugarcane Rules Committee*, ii, Patna, 1940.

16

Struggles in the Canefields: Small Growers in Mauritius 1921–1937

M.D. North–Coombes

Introduction

During the interwar years the international sugar economy was gripped by a severe crisis of overproduction, accumulation of surplus stocks, falling prices and reduced profits. The crisis had especially harmful repercussions in heavily export–orientated economies such as Mauritius where, by the early 1930s, sugar exports made up nearly 98 per cent of total export value.[1] Prices for Mauritian sugar fell precipitously and almost without interruption from a peak of Rs 50 in 1919 to Rs 5.90 in 1931, their lowest level in fifty years, and to Rs 5.50 in 1936.[2] Sharply reduced export earnings over such an extended period placed the Mauritian economy under severe pressure and it was claimed that 'the continuation of such low prices must mean the total collapse of the sugar industry on the island'.[3] However, although social tensions intensified in Mauritius during the interwar years and large–scale rural unrest erupted in 1937, the sugar industry did not collapse but instead recovered remarkably while social order was restored and the political and economic hegemony of sugar millers and large estate owners preserved.

It is argued in the paper that a proper appreciation of the overall impact of the sugar crisis in Mauritius must take into account the role of the small cane–growing sector, which had emerged in the island in the preceding three decades. An on–going transformation of the agrarian structure modified the impact of the sugar crisis and presented sugar millers with a wider range of responses to falling sugar prices on world markets than would have been possible within a less differentiated production system. Whereas sugar millers were able successfully to enlist the assistance of the colonial state in countering the major effects of the depression, small cane growers bore the brunt of this economic readjustment. The small growers suffered increasingly from official neglect and the exactions of sugar–millers, cane brokers and job contractors in their capacity both as cane farmers and field labourers on the sugar estates. Thus the actions of capital and the colonial state for the recuperation of profits in the sugar industry during the interwar years took place essentially at the expense of the immediate producers both in field and factory. The crisis also led to an intensification of the economic class struggle in Mauritius, and severely heightened social tensions in the island. Resistance by small cane growers and field labourers to continued wage–cuts and to arbitrary changes in the rates and methods of payment for cane culminated in widespread protests in 1937 when large–scale unrest broke out on the sugar estates.

This paper will be concerned with the differential impact of the sugar crisis on the small cane growers and field labourers in Mauritius. The effects of the sugar crisis on rural social conflict will also be investigated, with regards especially to the mass protests of 1937 and with the role of small cane growers in these protests. Finally, the broader significance of these rural struggles for the subsequent course of historical change in Mauritius and its implications for the small cane growers and field labourers will be tentatively discussed. It will be argued that although some gains were realized by the small cane–growers and casual labourers as a result of these struggles, these were more than offset by the effects of ruling class actions and strategies for the restoration of public order and social stability. A colonial counter–offensive sought to achieve this by combining repressive features with accommodative programmes of a regulatory and ideological nature. Their overall effect was to fragment and diffuse opposition to the *status quo* while guaranteeing the continued economic supremacy of the sugar millers.

The small cane–growing sector, the sugar crisis and the unrest of 1937

The importance of the small cane–growing sector

The small cane–growing sector as it emerged in Mauritius, consisted of petty cane farmers or *petits planteurs* as they were euphemistically called, who had access to land outside the sugar estates and of the *metayers*, to be found in the ranks of the labourers on monthly contracts, resident on the estates and holding small plots of land on a sharecropping basis. Small cane–growers already constituted an important component of the agrarian social structure by 1910, when indentured labour immigration was officially terminated, occupying nearly 36 per cent of the total sugar cane acreage.[4] Two–thirds of this area was situated off the estates and farmed by the *petits planteurs* either on a freehold basis or under a system of annual instalment payments over periods from five years upwards, after which title passed to the occupier. Instalments accrued to estate owners who had parcelled off portions of their properties (a practice known as *morcellement*) to raise funds in lean years or to finance agricultural or milling improvements. Not all small growers purchased land directly from mill or estate owners however. *Morcellement* led to an active land market where speculative transactions were rife. Insolvent estates were not infrequently bought by cane brokers and parcelled out for sale to petty cultivators, who undertook to grow sugar–cane exclusively for delivery to stipulated mills, either directly or through the intermediary of the cane broker.

There were also cane farmers resident on sugar estates who where known as *metayers*. The *metayers* were estate labourers who were allotted plots on the estates they resided on, with the stipulation that they would grow cane for delivery to the estate's mill. These labourers reportedly did not pay any money rent for the estate land they farmed, but they received less than the ruling market rate for the cane supplied, which was equivalent to payment of a concealed ground rent in kind. Moreover in boom years, the *metayers* were denied the opportunity of taking advantage of competition between millers for small growers' cane. Since the days of slavery and throughout the indentured labour period, estate workers had been granted access to garden plots which helped reduce labour costs to the planter, in providing for part of the daily reproduction of labour power. During the first world war, high sugar prices led to a shift to *metayage* on many estates, as provision grounds were placed under cane. The spread of *metayage* bound the labourers more closely to the factory owners by forging new chains of dependence, in the form of credit advances and purchases of cane. *Metayage* was however to decline sharply in importance during the subsequent sugar crisis.

The sugar–millers and estate owners had actively assisted in the establishment of a small cane growing sector in Mauritius, but they perceived the small–growers as a long–term threat to the monopoly over land and labour they had hitherto enjoyed. Hence it became a major area of ruling class concern and activity to guarantee the continued dominance of sugar monoculture over the economic landscape and to subordinate smallholder cultivation to large propertied and milling interests. With regards to the utilization of land, small cane growers were prevented from engaging in food crop production to any appreciable extent and were wedded to the cultivation of cane, through contractual agreements or the terms of land acquisition. On the eve of WWI cane was grown on 94 per cent of the cultivated acreage and the Royal Commission of 1909 remarked that food production was 'neglected to a degree unknown in other colonies'.[5] Small cane growers were almost entirely dependent on the market for their subsistence needs and the sale of their cane.

The chief advantages of a small grower class to the controlling interests in the sugar industry were as a flexible source of low cost cane supply and in its capacity as a reserve army of labour which could be drawn upon at will to meet the changing labour requirements of sugar production on a seasonal or casual basis. The small cane growing sector thus served both the cane supply and labour needs of the sugar industry. For both these requirements to be met it was essential that while large areas of land be made available for small–scale cultivation, individual access to land be limited. Official reports stressed the prosperity of small cane planters *as a group* by reference to their share of total cane cultivation, which was taken as evidence of Indian landed wealth in the colony. The contrasting reality of underemployment and grinding poverty, which was the lot of the individual smallholder in the island, was seldom explicitly acknowledged although evidence for it abounds in contemporary documents.

The first detailed estimates of the distribution of petty rural property in Mauritius were published in the agricultural censuses of 1930 and 1940. These illustrate starkly the preponderance of *minifundia* in the small–growing sector of Mauritian agriculture, where over 90 per cent of small cane–holdings off the estates were below 5 acres. The Census of 1940 showed that nearly two–thirds or more than 13000 smallholders in that category had access to plots that were smaller than one acre on average.[6] For the great majority of small growers cane farming on miniature plots was a hopelessly inadequate source of livelihood. There was too little land available for the small growers to subsist from the cultivation of their holdings and they were therefore forced into wage labour to supplement their income from cane farming. Moreover, the available evidence suggests that for most *petits planteurs* the proceeds from sugar cane sales to mills were less significant in the annual household budget than earnings from temporary wage employment on the sugar estates by the small growers and their families.[7] The representative *petit planteur* thus conformed more closely to Lenin's allotment worker than to the notion of an agrarian 'petty bourgeoisie', which Seegobin and Collen argue stood in conflict with the colonial or 'historic' bourgeoisie.[8]

This is not to say that the *petits planteurs* were a homogeneous social class. Over a thousand small growers worked plots larger than 5 acres according to the census of 1930, and more than two hundred of these were in effect capitalist farmers. They worked holdings between 20 and 100 acres, employing wage labour on a regular basis and adhering to agricultural practices which were not significantly different from those applied on sugar estates. The larger of these *petits planteurs* held a strategic place in the rural social structure. It was from this stratum that the middlemen and job–contractors on which the sugar millers depended for supplies of cane and casual labourers were drawn.

The small cane–growing sector on and off the estates also functioned as a source of flexible low cost cane supply to the sugar millers. By the early twentieth century estate cultivation had expanded to the limits of cultivable land, and it was not expected that the stock of land suitable for sugar cane cultivation could be significantly expanded by irrigation schemes or transport improvements. In these conditions, an increase in cane supply, which high sugar prices might call for, could best be met under existing circumstances by the extension of small–scale cultivation on to marginal land, both on and off the estates. This strategy held several advantages for the sugar industry. It meant that the risks associated with a sudden downturn in sugar prices and the consequent need for a contraction of the cane acreage would be carried by the small growers rather than by the estates. Secondly, it was felt that costs considerations of putting poor land under cultivation were less critical in the small growing sector which used family labour. It was believed that in that sector, production decisions were taken in line with the subsistence requirements of the household rather than in terms of the profit motive which regulated economic activity on the capitalist estates. How were the small cane growers affected by the sugar crisis?

Impact of the sugar crisis

In their capacity as cane farmers, the small cane–growers were subject to the exactions of the factory owners and of their intermediaries, the cane brokers and moneylenders. Falling sugar prices since the late 19th century, WWI excepted, gave sugar millers a constant stimulus to reduce the bargaining position of small cane growers with respect to cane purchases. As a result of a serious epizootic in 1902 marked by severe outbreaks of surra, an infectious animal disease deadly to cattle, horses and mules, the *petits planteurs* incurred large livestock losses and became dependent on estate tramways for the transport and sale of their cane to particular estates. However competition between millers for small growers' cane did not abate completely, and it was regularly claimed, in the face of falling sugar prices, that such competition was driving cane payment rates to unrealistically high levels. At the 1927 Mauritian Sugar Conference it was suggested that the millers follow the example of Trinidad where there was, 'the same destructive competition and that apparently this is being solved by a system of limitation of areas'.[9] This suggestion was readily taken up by the sugar millers. Factories agreed on regional boundaries for the purchase of the cane of small planters, whose bargaining position was correspondingly weakened. No controlling body existed in Mauritius, unlike the position in other sugar–producing regions such as Natal, to determine uniform rates of cane payment. These varied widely between different factories and suppliers of different standing. Arbitrary deductions from customarily accepted rates of payment were moreover widespread.

The *petits planteurs* conflicted with factory owners especially on the question of cane payment. They generally received a fixed quantity of sugar for each ton of cane delivered to the factory weighbridge, either in kind or in the form of a money equivalent, calculated on an average market price. Factory owners and managers in an endeavour to reduce costs of production complained that this accustomed rate was too high. At the Sugar Conference of 1927 it was suggested that payment to small growers be slashed from a customary 70–75 lbs to 60–64 lbs of sugar per ton of cane. The main justification given for the anticipated cuts was that the cane of small growers was of poor quality. The spread of chemical control in sugar–manufacture had made it theoretically possible to measure the sucrose content of cane, which, it was claimed, was substantially lower in the case of the cane of the *petits planteurs*.

Pro–miller commentators such as Sir F. Watts observed that the 'yield of cane obtained by these small cultivators is low, and it is estimated that while occupying 43 per cent of the area, they only produce 25 per cent of the crop'.[10] Though exaggerated, such observations were correct in substance. It is not surprising to find such a productivity differential between small cane–growers and estate land. Extremely short of capital, heavily burdened with debt, in insecure possession of micro–plots on marginal land and neglected by the colonial state in the provision of technical assistance and agricultural credit, small planters were unable to invest in agricultural improvements. This was in spite of their positive attitude towards innovation to which various commentators, not least of all Watts himself, testified.

In deploring the low productivity of small planter production, factory owners before 1937 did not yet feel compelled to adopt a more 'scientifically accurate' form of cane payment based on the richness of the cane measured in sucrose content. Arbitrary payment rates continued to prevail. Millers merely used the scanty scientific 'evidence' gathered by the technologists in their employ, as justification for intended unilateral reductions in payment rates, from higher levels negotiated in earlier times. Cane payment rates were subject to erratic changes. This arbitrariness was strongly resented by small cane–growers.

The uncertainty surrounding the price small planters would receive for their cane was compounded by the practice of compelling them to sell their canes to the factories through accredited cane dealers from whom they received reduced rates. Cane dealers performed the dual function of acting as intermediaries between factory owners and small planters in the marketing of crops and also as financiers supplying credit to the small growers. They directed cane loads to the weighbridges and tram lines according to milling requirements and at agreed rates. They were also the village usurers making advances on the security of land, crops or livestock, charging interest rates of 60 or even 120 per cent per annum. No alternative sources of credit were readily available to small planters except for the funds of co–operative societies set up after 1913. There were some 34 of these operating in 1938. With a membership of about 2000, they touched only a small fraction of the agricultural population.[11] The societies appear to have been dominated by a handful of wealthy growers who had invested relatively large sums in co–operative society shares. These men manipulated funds for their own benefit and even insisted on annual dividends being paid on their shareholdings. Mismanagement of the credit societies and the lack of adequate financial support from the colonial state, starved the small planters of credit. Thus the colonial state failed to cater for the agricultural credit needs of the small cane growers in any meaningful sense. This was in sharp contrast to the large loans granted to the sugar millers who benefited from numerous fiscal concessions and large colonial loans at preferential rates, which were extended to assist sugar millers in paying for technical improvements and to improve the industry's standing.

The partiality of the colonial state to the sugar oligarchy and its indifference to the small cane growers was also obvious in the field of technology. While it vigorously promoted technological change on the sugar estates, the colonial state did little to improve the technical efficiency of small scale farming. This lack was apparent in the absence of adequate extension services and in a failure to develop a suitably hardy cane variety to replace the deteriorating cane–types favoured by small–growers, such as the Uba. Technological improvements on large estates and in sugar mills by contrast proceeded apace with the aid of colonial funding and expertise. Mauritian sugar millers and estate owners did not conceal the fact that the search for technological improvements in the 1920s and 1930s was directed principally at reducing the wage bill in sugar cultivation and manufacture. Their strategy of responding to the sugar crisis through the adoption of cost–reducing innovations and mechanization had the full support of the authorities, strengthening the co–operation between state and capital in this field. For their part, estate owners formally called for the 'active co–operation and support' of the colonial government in seeking 'a better utilization of locally available labour.'[12] By 1932, the official almanac of the colony boasted that 'labour conditions were far less acute than a decade earlier, as a result of the widespread adoption of labour–saving appliances'.[13] Mechanization in field, factory and transportation allowed the same amount of work to be done by fewer hands. Attempts by the sugar oligarchy to force down wages and to reduce cane payments to small planters were facilitated by mechanization and technological innovations. These made capital temporarily less dependent on labour while making wage cuts possible. Improved chemical control in the factory and the increase in sugar yield per acre harvested on estate land moreover provided a justification for price discrimination against small planters in the purchase of their cane for milling and for arbitrary deductions on the grounds that this cane had a much lower sucrose content than estate cane.

The small cane growers also suffered from wage cuts and mounting unemployment or underemployment during the sugar crisis. Agricultural labourers in the sugar industry shared the common grievance of an insufficient income. Monthly labourers received Rs 10 a month plus rations while day labourers, paid by the task, earned Rs 30 per month. This contrasted sharply with the high salaries paid to estate overseers and managers who received between Rs 300 and Rs 750 per month plus quarters. With the onset of the sugar depression, as Kunwar Maharaj Singh reported in 1924: 'owners of estates were anxious, in view of the falling prices of sugar to reduce wages, or in the alternative to increase the hours of work.'[14] Similar observations can be found in almost every official report issued up to 1939. Wage cuts were facilitated in the 1930s by reductions in the prices of imported foodstuffs, but money wages tended to lag behind changes in the cost of living which, moreover, as the statistical indices show, fluctuated quite erratically.[15]

Agricultural labourers were the recurrent victims of a high rate of unemployment and underemployment during the eight months long inter-crop season, during which work on the sugar estates was all but unobtainable. With falling economic activity and retrenchment cyclical unemployment became an additional problem. At first it bore particularly harshly on the artisans in sugar factories. In time it effected small growers too. With the suspension of public works alternative avenues of employment to work on the sugar estates were closed. By February 1935 the Chamber of Agriculture alarmingly noted that unemployment in the island had reached 'disquieting proportions.'[16]

Casual agricultural labour was hired and supervised in their field tasks through the intermediary of job contractors or 'entrepreneurs'. These men recruited and deployed the agricultural labour force during the crop season, negotiated tasks and wage level and supervised field operations. This was a source of much dissatisfaction as the 'entrepreneurs' defrauded labourers of part of their earnings through concealment of rates contracted with the estates or the use of inaccurate measures for the assessment of tasks completed. This system became increasingly intolerable to the day labourers as economic conditions deteriorated.

The long period of depressed sugar prices after 1921 bore particularly sharply on the small growers, many of whom had purchased poor land at inflated prices from the estates during the war boom. Many fell increasingly into debt, some abandoned farming altogether or were simply driven off the land. It was reported in 1922 that several thousand acres planted under cane two years earlier had been abandoned with unfortunate results for the cultivators. By 1930 small cane–growers on and off the estates, cultivated only 54260 acres compared to 89153 acres seven years earlier. Their share of the total cane acreage fell from 45 per cent to 39.7 per cent in the same period – a proportion which continued to decline throughout the 1930s.[17]

The outbreak of 1937

As we have seen, small cane–growers clashed with factory owners on a variety of issues which concerned them both as cane suppliers to the central mills and as wage workers. Struggles over cane payment were noteworthy, playing a central role in the subsequent unrest. However, for the majority of small cane–growers the wage relationship represented a more fundamental contradiction. The question of wages was the focus of extensive political mobilization in 1936 and 1937, while wage demands were raised with great intensity during the unrest of 1937 and the job contractor system denounced.

Although there were isolated episodes of resistance against massive wage cuts, increasing unemployment, reduced rates of cane payment and other exactions, these failed to engender any concerted wave of opposition amongst the small cane–growers and agricultural workers of Mauritius. A report of 1931 remarked that 'an outstanding feature of the labour problem is the extraordinary adaptability and docility of the Indian labouring population. They have accepted the altered conditions due to the falling sugar market without murmur'.[18] The reasons for this comparative absence of labour militancy have been gone into in some detail elsewhere and will not be discussed here.[19] A continued deterioration in economic circumstances in the next five years set the scene for an unprecedented outbreak of popular militancy in rural areas. Increasing misery, in itself, would not necessarily have provoked the widespread unrest which engulfed the sugar estates in 1937. It was rather the combination of increasing misery with the emergence of a mass political movement, spearheaded by the Mauritius Labour Party in the second half of the 1930s in a changing international context, which broke through decades of passivity and unquestioning acceptance of even the grossest form of exploitation.

The immediate trigger of the unrest of sugar estates in Mauritius in 1937 was the arbitrary 15 per cent reduction in prices paid to small growers for the purchases of the Uba cane by Sans Souci and Rich Fund estates. The small growers decided to boycott the estates at Rich Fund as day labourers. They consequently stopped work and returned to their homes in the villages of Lalmatie and Union in the Flacq district on the east coast of the island. The strike swiftly spread to other estates which recruited labour or purchased cane from the Lalmatie area. Groups of strikers overturned cane trucks and stopped lorries ferrying labourers from reaching

Union–Flacq, Unite, Sans Souci and other Flacq sugar estates. Cane fields were repeatedly set on fire by the strikers. Groups of small growers and day labourers marched on estates and mills in the area to force them to cease operations and to demand redress from management for their grievances. On 13 August 1937 at Union–Flacq, which was owned by a prominent Indo–Mauritian family, the mill staff shot at a group of demonstrating workers. Four labourers were killed and six wounded. After the killings at Union–Flacq unrest broke out in other parts of the island. Through forceful protests and militant action the *petits planteurs* and labourers of Mauritius clearly indicated that they would no longer tolerate low prices in payment for their cane or low wages for their labour. The system of cane dealers and of job contractors was also strongly attacked and rejected.

The strikers were joined in their protests by sympathizers from outside the sugar economy. Police reports record the participation of stone masons, vegetable sellers, the unemployed and of a bread–seller, Codabaccus, who was killed by police at Camp La Hache in Souillac district. There is evidence of solidarity between small growers and labourers and within their ranks a unity which cut across ethnic divisions. Of particular relevance were reports of joint action by Indian and Creoles in Souillac against the police and estate owners.

The Mauritius Labour Party cannot claim direct credit for the widespread unrest which engulfed the sugar estates in 1937. The wave of popular protests, work stoppages and militant action which swept across the Mauritian countryside between 30 July and 28 September 1937 had a spontaneous character. The Labour Party played no role in planning or co–ordinating the activities of small growers and labourers. However, it is unlikely that the upheavals would have occurred on the same scale or with the same intensity without the period of political activism and rising popular consciousness which preceded it. Nevertheless, the protests lacked central direction and their thrust was consequently blunted and soon dissipated.

Although casual labour, recruited amongst the *petits planteurs*, shared many of the grievances of the resident monthly labour force on the sugar estates, the objective conditions of life and labour of these two groups of workers were quite dissimilar. Theses differences, which were accentuated rather than narrowed during the interwar sugar crisis, had a divisive effect on the agricultural labour force. Far from being an amorphous mass, the estate labour force in Mauritius was hierarchically stratified as well as structurally differentiated. An awareness of these divisions may assist our comprehension of the failure of 1937 unrest to cripple the sugar industry. Its impact was uneven and localized, and it failed to draw the support of the resident labour force on the sugar estates. Nearly six weeks after the unrest had broken out, small groups of monthly labourers went on strike at Labourdonnais and Sans Souci estates. At the end of September monthly workers also stopped work at the Beau Champ and Ferney estates. These strikes were peaceful and short–lived. They were brought to a rapid end by the offer of limited concessions by the estate managers concerned. Promises were made of end–crop bonuses, with or without wage increases, and of improvements in the quality of rations. There is nothing to suggest, however, that the monthly labourers supported the objectives of the small growers and day labourers or that more than a small proportion of the resident labour force was involved in these isolated protests. The scale and intensity of the struggles of the villagers against the millers, estate owners and police, on the other hand, necessitated far more than *ad hoc* solutions.

The historical significance of the 1937 unrest on the sugar estates

The initial response of the colonial state and of the dominant sugar interests to militant worker action and political activity during the late 1930s was one of panic and repression. The Chamber of Agriculture blamed the unrest on agitators and formed a 'Vigilance Committee' which met behind closed doors, to keep the labour situation under surveillance and to plan appropriate responses to labour action.[20] After the unrest had died down, a new governor, Sir Bede Clifford, took further steps to preserve what he called 'public security' by adopting extended police measures. Clifford cracked down on political activists and deported the labour organizer, Anquetil, to Rodrigues island.

Clifford presided over the implementation of some of the proposals put forward to resolve the crisis of 1937 by the Hooper Commission of Enquiry into the unrest.[21] The Commission had no small planter of labour members, or even a single representative from the Indo–Mauritian community. It failed to collect much evidence from persons actually involved in the strike. The Commission nevertheless addressed itself directly to the main grievances of the small cane growers and agricultural labourers. Its recommendations have been represented in Mauritian historiography as important concessions granted to the rural majority. The strikers are portrayed as having achieved concrete gains from their militant activity and sacrifices.[22] What is not acknowledged in mainstream historical writing on modern Mauritius, however, is the extent to which the reformist legislation flowing from the Hooper Report was designed to prevent a recurrence of 1937 for all time.

The reforms adopted by Clifford had the effect of creating an organic interdependence between the various components of the agricultural body. Organizational structures were erected for the regulation of conflict between millers and *petits planteurs*, over which the colonial state presided, ostensibly as neutral arbiter.[23] In the resolution of conflict, appeal was now to be made to the wisdom of the law or to that of the 'science' of sugar technology, which were implicitly defined as standing above conflict. The sugar millers, who had a monopoly of access to scientific and legal resources, thus stood immeasurably reinforced in struggles with the *petits planteurs*. The legal establishment in Mauritius was dominated by scions of the Franco–Mauritian elite. The Mauritian College of Agriculture, where sugar technologists were trained for the island's factories, for decades had no Indo–Mauritian graduates. When unrest flared up amongst monthly labourers in 1943, the estate owners involved insisted on communicating with the strikers and even the police, through the medium, or in the presence, of their legal representatives.[24] The resolution of conflicts was henceforth to be decided in the courts rather than in the cane fields. Within these new parameters, property and contractual rights were strictly enforced and the scope for mounting radical attacks on the existing order, narrowly circumscribed. Cane payment became a subject of endless and unsuccessful litigation. However so long as cane quality continued to differ on the holdings of small growers and large estates the struggles over payment shares according to scientific principles merely obscured the reality of disproportionate rewards earned by large landowners by virtue of their 'superior efficiency'.

In their dealings with small cane growers millers made increasing appeals to expert opinion. Once the initial principle of payment according to sucrose content was accepted, disputes about the small growers' share could be reduced to dispassionate scientific questions. When a leading article appeared in the *Mauritius Times* in January 1962, attacking the treatment of small growers by sugar millers, the factory owners obtained the appointment of two British technical experts to investigate the allegations of the article, 'The Fraud Continues'. The Balogh Report, published under government auspices, slated the *Times* article for its unscientific quality. It recommended that the sugar industry take steps to 'dispel this technical ignorance' and to demonstrate to the public 'why certain things are being done'. The report continued:

> To promote more constructive attitudes we have incorporated in our report two chapters which contain a considerable volume of technical details. We have done this so that the public in Mauritius may know something of the history leading up to the present method of calculating how the quantity of sugar produced from each planter's cane is assessed, and may also be aware of the precise formulae which are applied in this calculation.[25]

The Balogh report illustrates clearly how a ruling class has the power to rewrite its own history, even after a short space of time, and emerge strengthened from what may have appeared initially as a defeat.

The Clifford Reforms improved the legal position and status of rural workers in Mauritius, while the bargaining power of both rural workers and small cane growers was increased in their dealings with the sugar estates and mills. Moreover, methods were devised to regulate conflict by drawing it into particular channels, while it became possible for the small planter elite and the more pliable worker leadership to be incorporated more firmly in the establishment. A system of collective bargaining, loaded in favour of employers, was created for the sugar industry and a Department of Labour founded. Tame trade unionists were brought out from England to 'school' Mauritian labour leaders in the proper practices of labour organizations.

Only in one important respect did 1937 open up new opportunities for class struggles in Mauritius. The labour unrest modified the role of the colonial state and created a rift between metropolitan authority and the largely Franco–Mauritian colonial bourgeoisie, The latter resented the reforms, felt betrayed and unjustly accused of ill–treating their labourers.[26] For the colonial state it became necessary to broaden the social basis of support for colonial rule, and constitutional reform was placed firmly on the agenda, much against the wishes of the sugar oligarchy. These cracks in the facade of colonial rule were strategically exploited by the dominated groups in the colonial social formation.

Thus 1937 marks, in many ways, the beginning of a long phase of struggle for political reforms which culminated in the new constitution of 1948, the granting of universal franchise in 1959 and of formal independence in 1968. The small cane growers and the agricultural labourers, in alliance with sections of the urban petty bourgeoisie, were to provide the leadership and mass support for the drive against the continued power and privileges of the sugar oligarchy. In retrospect this struggle for political change from the mid–1930s onwards can be seen as a prerequisite for greater economic and social progress. However, because it failed to address itself to the question of fundamentally transforming the economic system, it neglected to attack the material basis for the continued dominance of sugar millers and large estate owners, which remained unbroken. Although the sugar oligarchy ultimately lost its political pre–eminence, it continued to enjoy economic pre–dominance. Moreover, the colonial counter–offensive had blunted the edge of the anti–colonial challenge and confirmed its reformist nature. As was seen in the case of the small cane growers, a mix of repressive and accommodative programmes after 1937 served to co–opt segments of the opposition forces and to stabilize the colonial order. In this objective it may be said to have succeeded beyond measure.

References

1. Imperial Economic Committee, *Plantations Crops: A Summary of Figures of Production and Trade relating to Sugar, Tea, Coffee, Cocoa, Spices, Tobacco and Rubber*, London, 1936, 15.

2. A. Bax, ed. *Mauritius Almanach and Commercial Handbook for 1939–41*, Port Louis, 1941, E37.

3. A. Bax, ed. *Mauritius Almanach and Commercial Handbook for 1930–31*, Port Louis, 1931.

4. M. Koenig , *Agricultural Census in Mauritius, 1930*, reprinted in *Revue Agricole de L'île Maurice*, Nos. 56–58 (March–April–July–August 1931).

5. Parliamentary Papers, United Kingdom, *Report of the Mauritius Royal Commission, 1909*, June 1910, Cd.5185, 18.

6. Colony of Mauritius, *Report on the Agricultural Census, 1940*, by M. Koenig, Port Louis, 1940.

7. See the estimate of hypothetical household budgets in Parliamentary Papers, United Kingdom, *Financial Situation of Mauritius, December 1931*, March 1932, Cd.5034.

8. R. Seegobin and L. Collen, 'Mauritius: Class Forces and Political Power', *Review of African Political Economy*, 8 (Jan–April 1977), 110–112.

9. The Mauritian Sugar Industry Conference 1927, *Resolutions and Memoranda*, Port Louis, 1928.

10. Parliamentary Papers, United Kingdom, *Report on the Mauritius Sugar Industry by Sir Francis Watts, 1929*, March 1930, Cd.3518, 33.

11. Colony of Mauritius, *Annual Report on the Co–operative Credit Societies for 1938*, Port Louis, 1938.

12. Mauritius Chamber of Agriculture, Minutes of Meetings.

13. A. Bax, ed., *Mauritius Almanach and Commercial Handbook for 1932–33*, Port Louis, 1933, 27.

14. Government of India, *Report by Kunwar Maharaj Singh on his Deputation to Mauritius*, 1925, Delhi, 1925, 22.

15. A.D. Britter, *A Commentary on Facts: Being a Survey of the Principal Issues Raised by the Recent Unrest on the Sugar Estates of Mauritius*, Port Louis, 1937, 35, 38 refers to the role of falling imported commodity prices in accounting for 'the general reduction of costs of production between 1929 and 1936'.

16. Mauritius Chamber of Agriculture, Minutes of Meeting, reprinted in *Revue Agricole de L'île Maurice*, No. 79 (Jan.–Feb. 1935), 13.

17. M. Koenig, *Agricultural Census*, 1930.

18. A. Bax, ed., *Mauritius Almanach for 1930–31*.

19. These are discussed to some extent in a fuller version of this paper. See, M.D. North–Coombes, 'Struggles in the Cane Fields: Small Cane Growers, Millers and the Colonial State in Mauritius, 1921–37', *Journal of Mauritian Studies*, 2, 1, (1987).

20. Mauritius Chamber of Agriculture, Minutes of Meeting held on 3rd September 1937, 3–4.

21. Colony of Mauritius, *Report of the Commission of Enquiry into Unrest on Sugar Estates in Mauritius*, 1937 by C. A. Hooper *et al.*, ort Louis, 1938, 140.

22. See for instance K. Hazareesingh, *History of Indians in Mauritius*, London, 1975 and N.M. Varma, *Mauritius on the Move*, Quatre Bornes, 1979.

23. 'Struggles in the Cane Fields', *Journal of Mauritian Studies*, (forthcoming).

24. Colony of Mauritius, *Report of the Commission of Enquiry into the Disturbances which Occured in the North of Mauritius in 1943*, by S. Moody *et al.*, Port Louis, 1944.

25. Mauritius Legislative Council, *Commission of Inquiry (Sugar Industry) 1962* , by T. Balogh and C.J.M. Bennet, Port Louis, 1963.

26. See for instance A.D. Britter, *A Commentary on Facts*.

17

Protectionism and Sugar Production in Central and Equatorial Africa, 1910–1945

Gervase Clarence–Smith

The sugar industries of Central and Equatorial Africa were doing rather well overall in the inter–war years.[1] In 1911, the region accounted for 0.4 per cent of the cane sugar produced in the world, a proportion which had risen to 0.6 per cent by 1939.[2] In Angola, the Belgian Congo (Zaire) and Southern Rhodesia (Zimbabwe), the industries grew from a low or non–existent base during the 1930s, at a time when sugar producers elsewhere were reeling under the hammer blows of the recession. In contrast, the older and larger sugar industry of Mozambique suffered considerably during the recession. The key to this difference lay in the relation between protection and production. The internal markets of the four sugar producing colonies were too small to warrant the development of modern sugar industries. With the exception of Rhodesia, the key to prosperity thus lay in access to protected metropolitan markets in Europe.

If the sugar industries of Central and Equatorial Africa were marginal to the world sugar economy, they were vital to Angola and Mozambique. By 1943, there were in the two colonies some 41,000 African labourers and 1,000 European staff employed producing 135,000 tons of sugar from cane planted on over 30,000 hectares. Total investment was said to be in the order of seven million pounds sterling. In Portugal itself, nearly 5 per cent of ordinary state revenue came from taxes and duties on sugar, almost all of which was imported from Angola and Mozambique. It was the expanding sugar sector which carried the storm–beaten European planters of Angola through the 1930s recession, at a time when coffee and sisal markets had collapsed. As for Mozambique, it was sometimes said, with some exaggeration, that 'the colony is sugar'. By the mid–1940s, sugar accounted for 29 per cent of the capital invested in land and plant, 13 per cent of the cultivated area, and 12 per cent of the colony's foreign exchange earnings.[3] In contrast, the sugar industries of the Congo and Rhodesia were late–comers, and far less significant in their respective economies. Thus, in 1944, there were only 140 hectares under cane in Rhodesia, and production was 1,000 tons or less. Congolese production was higher, at some 13,000 tons, but this was still only a tenth of the output of the Portuguese territories.[4]

The political economy of sugar procurement in Portugal

The Angolan and Mozambican sugar industries had been turned towards the distillation of rum until this was banned in stages from the turn of the century, and they faced problems converting to sugar. Each colony was guaranteed a 6,000 ton import quota into Portugal at a standard 50 per cent rebate on import duties, but the treasury was reluctant to go above this, for fear of losing the revenue from high import duties on foreign sugar.[5] Parliament thus refused to approve a bill in 1912, which would have allowed the importation of all colonial sugar at the 50 per cent rebate. However, two years later, the 6,000 ton quota was guaranteed until 1933, and Angola and Mozambique were both allowed to increase it by 600 tons a year up to a total of 18,000 tons. The principle of equality between the two territories was thus tacitly abandoned in favour of Mozambique, whose production was already around 30,000 tons, at a time when Angola was struggling to produce 4,000 tons. The Cape Verde islands were also allocated a quota of 1,000 tons a year, but exports from these arid islands averaged a mere 3 tons a year. At this time, Portugal's annual sugar consumption was some 35,000 tons.[6]

The situation was further complicated by the question of sugar production in the Azores and Madeira. Both archipelagoes were in law integral parts of Portugal, but quotas on Azorean beet sugar exports to continental Portugal had been enforced since 1903, and Azorean sugar had to pay the same duty as that placed on colonial sugar. In contrast, the British company which held a monopoly of cane sugar production in Madeira insisted on its historic right to unimpeded and untaxed access to the Portuguese market. In return, it accepted a new contract with the government in 1911, in terms of which it had to pay Madeiran peasants the astronomically high sum of £3/10/– sterling for a ton of cane. But the overall capacity of these two small territories was limited, and their exports to Portugal, though growing, only came to some 6,500 tons in 1914.[7]

A much greater threat to colonial planters came from the pressures to develop beet cultivation in continental Portugal, copying the successful example of neighbouring Spain. After the 1910 republican revolution, treasury objections to loss of revenue from import duties had to be balanced against newly powerful rural interests, who argued that beet not only provided a valuable cash crop for the depressed agrarian sector, but also improved crop rotations and provided fodder for cattle. Moreover, beet sugar cultivation could bring down the cost of living for the urban petty bourgeoisie, who strongly supported the new régime. Experimental beet cultivation thus began in continental Portugal some time before 1918. The following year a decree legalized commercial exploitation of beet sugar, offering various incentives. In spite of a puzzling lack of success, the legalization of beet cultivation was considered a sword of Damocles by colonial planters.[8]

However, during and just after the first world war, the Portuguese market ceased to interest colonial producers, as world prices soared above those in Lisbon. Prices were pushing £100 sterling per metric ton in London, when the official Lisbon customs valuation for raw sugar was between £20 and £25. Inevitably, Mozambican and Angolan sugar began to gravitate towards the international market.[9]

Matters became more complex as sugar prices climbed steeply in Lisbon, with the official customs valuation for 1920 catching up with London prices, still over £70 a metric ton. The republican régime's very survival was threatened by runaway inflation and it reacted by trying to hold down prices. A decree of 1920 obliged the colonies to deliver fixed minimum quantities of sugar at prices well below those prevailing on the world market. In 1920, a quota of 16,700 tons was demanded from Mozambique, raised to 30,000 a year later. Angola was asked for 3,300 tons and then for 6,000. The government fixed a price at roughly £11 a ton, which fell to nearer £6 by 1922 with the depreciation of the *escudo*, at a time when London prices were still around £20 a ton. However, the republic was unable to enforce these measures effectively, partly because of growing political chaos internally, but also because of an acute lack of Portuguese shipping, which was alone authorized by law to carry colonial raw materials to Lisbon. The sullen passive resistance of the colonial sugar companies made things no easier. In the event, Mozambique sent only about half its quota to the metropolis in 1922. By 1925, the whole scheme had been quietly shelved. Lisbon prices were back roughly in line with London, and Mozambique was at last sending about 30,000 tons to Portugal. As volatile sugar prices continued to plunge on the world market, and subsidized German beet sugar began to make significant inroads in Portugal, colonial producers once again clamoured for protection on the Portuguese market. A decree of 1925 raised duties on foreign sugar, but more as a revenue measure than as effective protection for colonial producers.[10]

In 1926, the tottering republican régime was overthrown by right–wing army officers, who were much more responsive to the demands of the planters of Angola and Mozambique. In 1927–1928, beet cultivation in continental Portugal was banned.[11] In 1928, the Azores were prohibited from exporting sugar to continental Portugal, and this was extended to Madeira in 1934, although temporary concessions for small amounts of sugar were granted during the depression years. Some 10,000 poor peasants in the Atlantic islands, as well as the unpopular British sugar mill in Madeira,were thus sacrificed to a handful of rich Portuguese settlers.[12] At the same time, greater colonial sugar imports into Portugal were secured by raising duties on foreign sugar in 1927 and allocating a 77,000 ton quota to the colonies under the standard 50 per cent customs rebate. The lion's share of the quota went to Mozambique, with 62,000 tons, whereas Angola was given only 14,000 tons. The remaining 1,000 tons were left to the Cape Verdes as before.[13]

However, when Salazar first rose to political prominence as finance minister, he decided that the terms were financially unacceptable for the treasury, as well as being too favourable to the foreign dominated Mozambican industry. The quotas were therefore readjusted in 1928, with a total of only 65,000 tons allowed in under the 50 per cent rebate, of which Angola now obtained 25,000 tons and Mozambique a mere 40,000 tons. As Angola was unable to meet this quota, whereas Mozambique's total production was nearly 100,000 tons, this was seen as a gross injustice by Mozambican producers. A 'national salvation tax' was also imposed on all sugar purchased in Portugal, causing consumption to drop. Although it is hard to disentangle the effects of the recession from those of higher prices, it is certain that Portuguese consumption fell from over 85,000 tons in 1928 to 63,500 tons in 1933. The sugar producers pointed out that in 1938 the Portuguese consumed 9.4 kilos of sugar per head, compared to 50.6 in Britain.[14] In the mid–1930s, one of the Angolan companies calculated that 46.3 per cent of the price paid by Portuguese consumers went straight into the public coffers.[15]

As world conditions worsened and prices plummeted, Salazar took new measures to save the planters and prevent the dumping of foreign sugar. A decree of June 1930 created a sliding scale tax, which automatically made foreign sugar more expensive to the consumer than colonial sugar. The price of colonial raw sugar was fixed at around £15 sterling a metric ton, at a time when world prices reached a low of under £5 in 1934–6. The effects of the 1930 measures were plain to see, for the colonial share of the Portuguese sugar market never fell below 90 per cent during the 1930s, compared to an average 59 per cent for the previous decade. This satisfied the planters' deeply held belief that the Portuguese market should be reserved for them at a 'fair' price. However, there were aspects of the 1930 legislation which were less to the liking of the 'sugar barons'. The 1,000 ton quota for the Cape Verdes, at a time when the islands ceased exporting sugar altogether, and the unrealistic equal quota for Mozambique and Angola allowed the government to continue extracting high duties on supplies imported from abroad to make up for some of the shortfall.[16]

Colonial sugar producers also chafed at Salazar's policies on sugar refining. Imports of refined white sugar from the colonies into Portugal had traditionally been forbidden, to protect the refining industry in the metropolis. The colonial companies got around the ban in the 1920s by creating or buying their own refineries in Portugal. But relations between local refineries and those owned by colonial concerns became fraught, as refining capacity considerably exceeded the reduced consumption of the early 1930s. In 1934 the state therefore intervened to protect the smaller less efficient refineries, by fixing quotas under its corporatist policy of industrial regulation. The Refinaria Colonial, belonging to the Sena Sugar Estates of Mozambique, was alone able to produce over 60,000 tons a year by the early 1940s, had it been allowed to work six days a week and 24 hours a day. The refineries belonging to other colonial firms could have furnished another 40,000 tons, and this at a time when Portuguese consumption had risen again to about 70,000 tons. But government quotas reserved a third of the market for the small, old, inefficient, independent refineries of the metropolis.[17]

Salazar's 1932 decree on alcohol production was also met with mixed feelings. Inspired by the Brazilian example, this was intended to save foreign exchange on petrol imports and to prevent the companies dumping their molasses into the nearest river, because the market for molasses in South Africa and Britain had disappeared with the onset of the recession. Some industrial alcohol had been distilled for pharmaceutical and domestic purposes since the republicans passed enabling legislation in the 1910s, but the market was small and uncertain. In 1932, the state guaranteed a fixed price for industrial alcohol at 1.20 *escudos* (c. 2s.5d) a litre. The alcohol was then mixed into a fuel which included three parts of petrol for one of alcohol. However, the price fixed for this mixture was too high for it to sell really well. By the early 1940s, Mozambique had a capacity for distilling some six million litres of industrial alcohol, but production was only about one and a half million litres, and the new fuel was only widely used in the centre of the colony. Angola's production was about the same in 1945. The monopolistic nature of the concessions was also unpopular with those companies who were left out in the cold.[18]

The outbreak of the second world war created new sources of dissatisfaction for the 'sugar barons'. The price paid by neutral countries for sugar rocketed to around £50 sterling a ton, while British controlled base prices were set at around £18 a metric ton. Salazar kept the fixed price in Lisbon more or less in line with British base prices, ignoring both the high free market prices and the anguished protests of colonial producers, concerned by steeply rising production costs.[19] Some sugar was exported to neutrals, especially Switzerland, but the British were ruthless in preventing this trade from becoming a camouflaged method of supplying the Axis powers.[20]

213

The evolution of sugar production in Africa

Mozambique's sugar industry was older and larger than that of Angola and it grew rapidly up to 1929, under the stimulus of high world prices, followed by an increasing share of the quota for the Portuguese market. From the mid–1920s, Portugal regularly took over half of its East African colony's output, which rose from 43,000 metric tons in 1917 to a peak of 91,000 in 1929.[21]

Regional markets helped to boost production. South Africa was initially the most promising client, for an agreement in 1909 allowed Mozambican products into South Africa free of duty. South African prices were high,and transport costs were low, making this an attractive market. In return for this and other concessions, the Portuguese allowed labour recruitment in Mozambique for the South African mines. However, the Natal sugar interests in South Africa were resolutely opposed to this arrangement and they pressed hard for its revocation. They were successful in 1923, when an import tax of £4/10/– (S.African) a ton was placed on Mozambican sugar, raised to £8 in 1926. Under these conditions, Mozambican sugar exports to South Africa, which had reached 12,000 metric tons in 1922, found it harder to compete with Natal produce. However, South Africa still took over 10,000 tons in a good year like 1928.[22] Moreover, the Mozambican planters found a partial substitute for South Africa in British Central Africa, which had no sugar industry of its own. The first exports were sent in 1924, and Mozambique obtained a commanding position in a market which accounted for some 10,000 tons a year in the 1930s.[23]

The crisis came as world prices fell and Portugal decided to restore parity in the quotas allocated to Angola and Mozambique. Although parity took a decade to achieve, Mozambique's share of the protected home market was slowly but surely pushed down from a little under 50,000 tons in 1929 to a little over 30,000 tons a decade later. At the same time, the South African market was completely closed in 1932, when the protective tariff was doubled.[24]

There was thus a scramble for alternative markets. Small quantities of sugar were placed around the Indian Ocean, where Mozambique enjoyed transport advantages but faced formidable competition from Mauritius and South Africa. The planters encouraged the habit of drinking tea in order to stimulate local consumption of sugar, which rose from some 4,000 tons in 1928 to over 20,000 tons by 1944. The international sugar agreement of 1937 gave the colony an unexpected boost, allocating it a 30,000 ton quota on the world market. During the second world war, the Mozambican sugar industry enjoyed low priority in Allied allocation of shipping space, but demand increased, lucrative neutral markets emerged, and Rhodesian competition in Central Africa failed to develop.[25]

Angola's experience was almost the mirror image of that of Mozambique. Development was slow and faltering, until equality in quotas on the Lisbon market was reintroduced in 1928. Portuguese imports from Angola then went from under 10,000 metric tons in the mid–1920s to some 40,000 tons in the early 1940s. The internal market also grew, though less than in Mozambique. Stringent exchange control measures adopted in 1931 cut imports of refined sugar from some 500 tons a year to practically nothing and internal consumption rose from about 2,500 tons a year in the early 1930s to some 7,000 tons in the late 1940s, with total production reaching just over 50,000 tons by 1946.[26]

The sugar industry of the Belgian Congo was a product of the euphoria on sugar markets after 1918. In October 1923, a company obtained a regional monopoly on sugar milling for twenty years in the Kwilu region, but the first 800 tons of sugar were not produced till 1929, when the world sugar economy was already in deep recession. The local market was secure, with imports steadily reduced and eliminated by the early 1940s, but local consumption was only 1,200 tons in the late 1920s, 2,800 tons in 1939, and 8,000 tons in 1945.[27] The company enjoyed tariff–free access to French Equatorial Africa, owing to the provisions of the Congo Free Trade Zone, but consumption there was a mere 600 tons a year in the mid–1930s.[28] Other parts of the Zone were cut off by high transport costs.

The Belgian market thus became the company's refuge in the 1930s. Legislation guaranteeing access to the metropolitan market was passed in 1924, and as the economic crisis of the 1930s deepened, the Congolese authorities suspended export duties. This gave rise to protests by Belgian beet producers. However, with Belgian consumption at around 250,000 tons a year, colonial cane sugar posed only a minor threat to the beet growers. A compromise was reached, imposing maximum quotas on imports of colonial cane sugar.[29] Under this arrangement, production in the Belgian Congo rose fairly steadily from 1,000 metric tons in 1930 to 15,000 metric tons in 1939. New difficulties emerged when Belgium was occupied by Germany in 1940, and production stagnated at around 13,000 tons for the rest of the war.[30]

For Rhodesia, the existence of major sugar producers in the British empire made any idea of exporting to Britain utopian. The home market was small, and the autonomous settler government was hostile to forms of protection which put up the cost of living and the cost of labour. However, the recession led to a greater interest in manufacturing as a panacea for white unemployment, and war–time supply difficulties gave a further boost to industry. The sugar industry of Rhodesia thus began with refining rather than planting, in clear contrast to the other cases in the region. Before the recession, Rhodesia imported all its sugar in the form of refined white sugar, mainly from Mozambique. Tariffs were thus imposed on refined sugar, while rebates were granted for imports of raw sugar. A refinery was set up in 1935 in Bulawayo, and a year later it passed into the hands of Mozambique's Sena Sugar Company.

Sugar planting progressed much more slowly under these conditions, although experimental planting took place from 1931 in the Lowveld region close to Mozambique. But technical and financial problems beset the enterprise. The first tons of raw sugar were not milled till 1937 and production was still only about 1,000 tons in 1944. An impatient government decided to buy out the enterprise a year later.[31]

Protectionism versus comparative advantage

The hardest question to answer is whether these sugar industries would have enjoyed any comparative advantage in a free market, or whether they could only have grown under the shadow of protectionism. The sugar companies took the line that they would have found it difficult to survive on a free world sugar market. Opponents of the companies argued that African sugar was exceedingly cheap, and that the companies were making 'super–profits' behind protectionist barriers. One immediate observation is that no Mozambican sugar company paid a dividend in the period 1925 to 1935, and that they probably did not do so for the rest of the 1930s.[32] It is true that the Companhia do Açúcar de Angola paid high dividends throughout the recession, but this could simply have reflected the high degree of protection given to the Angolan industry.[33]

From some points of view these sugar enterprises were not particularly favoured. The lands of Central and Equatorial Africa were poor compared to those of other tropical sugar producers, although they were cheap to buy. Irrigation was a necessity and the unpredictable rivers of Africa were liable to burst their banks, in spite of expensive dykes. The producers also suffered from plagues of locusts which were particularly severe in this part of Africa in the 1930s. Moreover, the small size of all these industries made it difficult for them to benefit from economies of scale and the newest technologies. And the companies all suffered from the acute lack of development of the economies they operated in, in terms of transport, banking, insurance, maintenance facilities, and other services, all of which pushed up costs.[34]

The modernity of these industries varied greatly, and this is a subject which requires more research. Saldanha in the late 1920s considered that the Mozambican industry as a whole was characterized by archaic production methods and ancient machinery, but he was a highly prejudiced settler with a bitter personal grievance against the companies.[35] In fact, there existed a mixture of new and old, with some second–hand plant and some at the forefront of new technology.[36] Central milling integrated with plantations remained the rule, with the partial exception of one Mozambican company which bought a proportion of its cane from small white farmers.[37] Sena Sugar's only labour saving devices in the fields were steam ploughs and locomotives to drag the cane to the factory, although ridging machinery was introduced in the late 1930s. It was not until 1943 that the company began to consider the problem of soil conservation and adopted a programme of green manuring.[38] The companies also tended to cling to the Uba cane, hardy and resistant, but with poor yields.[39] However, the Congolese plantations experimented with high yielding P.O.J.2878 from Java from the outset, and it was introduced in Angola in 1931.[40] Moreover, the sugar companies were usually the most highly mechanized and modernized agricultural producers in the local economy. And labour intensive methods often made most sense when labour was abundant, capital scarce and expensive, and infrastructures poor or non–existent.

This raises the trickiest subject of all, namely labour costs. Detractors of the companies, from both left and right political perspectives, argued that African labour was extremely cheap in world terms, and that this was the chief comparative advantage enjoyed by the sugar producers. As was generally the case in the sub–continent, the bulk of the labour force was composed of single male migrants. Support for the families of migrants,and for the labourers themselves in their periods at home, did not come out of wages, but was borne by the agricultural sector in their places of origin. On top of this, both the Portuguese and the Belgian empires were notorious for their forced labour practices, which further depressed wage levels. It was thus estimated that wages in Central Mozambique were among the lowest in the world.[41]

The problem with this 'cheap labour' argument is that it fails to come to grips with the real costs of labour to employers, which were determined as much by productivity and overhead recruitment costs as by wages. The companies were far from seeing their labour as cheap. Indeed, they looked enviously at the favourable wage and productivity combination existing in Java. Recruitment costs for migrant labour were higher than for a stable labour force, and the insecurity was greater. But the real problem was the abysmally low productivity of African labour. Lack of the most basic skills, short periods of employment, frequent absenteeism and desertion, poor motivation and malnutrition combined to make labour productivity a constant headache for employers.[42] By the 1920s, some of the companies had begun to realize that low wages did not just result from low productivity, but actually locked the labour system into a vicious circle, and they made some attempts to reform their labour practices.[43]

If the competitiveness of these sugar industries remains an open question, it is certain that in the real world of 1914–1945 there was no possibility for them other than protection. The increasingly autarkic world economic climate developing from 1914 made any other solution for the sugar industries of Central and Equatorial Africa illusory. And as the world returned to free trade after 1945, the sugar industries of Central and Equatorial Africa reacted by gradually retreating from exports, turning to the rapidly growing internal markets within the continent.[44]

References

1. A longer version of this chapter is appearing in Portuguese in the *Revista Internacional de Estudos Africanos*, containing more information on the sugar companies.

2. World production & 1930s Angola & Mozambique figures: Sena Sugar Estates Ltd., *Moçambique e o problema açucareiro*, Lisbon, 1945, 10–11; Mozambique 1911: R. Lyne, *Mozambique, its Agricultural Development*, London, 1913, 53; Angola 1910–1925: *Boletim da Agência Geral das Colónias*, III, 22 (1927), 121; Belgian Congo: Belgium, Ministère des Colonies, *Plan décennal pour le développement économique et social du Congo Belge*, Brussels, 1949, II, 575; Rhodesia: Federation of Rhodesia & Nyasaland, *Report of the Commission of Enquiry into the Sugar Industry*, Salisbury, 1962.

3. Sena Sugar, *Moçambique*, 114–16,119.

4. Belgium, *Plan décennal*; Federation of Rhodesia & Nyasaland, *Report*.

5. W.G. Clarence–Smith, 'The Sugar and Rum Industries in the Portuguese Empire, 1850–1914', in B. Albert and A. Graves, eds., *Crisis and Change in the International Sugar Economy, 1860–1914*, Norwich & Edinburgh, 1984, 227–35.

6. *A nova questão Hinton*, Lisbon, 1915, 27–8; Sena Sugar, *Moçambique*, 56; A. Carreira, *Estudos de economia caboverdiana*, Lisbon, 1982, 254–55.

7. *A nova questão Hinton*; Clarence–Smith, 'The Sugar and Rum Industries', 231–33.

8. H. G. de Amorim Parreira, 'Historia do açúcar em Portugal', *Anais da Junta de Investigações do Ultramar*, VII, 1 (1952), 216–7; A.A.Lisboa de Lima, 'O problema do abastecimento do açúcar e do seu barateamento', *Boletim da Sociedade de Geografia de Lisboa*, XXXIV (1916), 266–7; A. F. de Assis Andrade, *Portugal economico; teorias e factos*, Coimbra, 2nd edn., 1918, 175–84.

9. E. de Almeida Saldanha, *Mais uma burla dos açucareiros, carburante colonial e carburante nacional*, Oporto, 1933, 14; Sena Sugar,*Moçambique*, 19–20.

10. Sena Sugar, *Moçambique*, 20–2, 57–8; Parreira, 'Historia do açúcar', 223.

11. A. Castro, *O sistema colonial português em Africa*, Lisbon, 2nd edn.,1978, 132, note.

12. Parreira, 'História do açúcar', 245–6, 239–40; L. de Sousa Lara,'A Indústria do açúcar na economia do imperio', *Boletim da Sociedade de Geografia de Lisboa*, LIX (1936), 173.

13. Sena Sugar, *Moçambique*, 22–4.

14. *Ibid.*, 24–5, 136–43.

15. Lara, 'A indústria', 174.

16. Sena Sugar, *Moçambique*, 41–50, 130, passim.

17. Sena Sugar, *Moçambique*, 165–72; Parreira, 'História do açúcar', 271, 281–82.

18. Saldanha, *Mais uma burla*; Sena Sugar, *Moçambique*, 160–62; A. A.Diogo, 'Evolução industrial de Angola – o açúcar', *Actividade Economica de Angola*, 42–3 (1955).

19. Sena Sugar, *Moçambique*, 94–6, 130.

20. G. Clarence–Smith, 'The Impact of the Spanish Civil War and the Second World War on Portuguese and Spanish Africa', *Journal of African History*, XXVI, 4 (1985), 309–26; Diogo, 'Evolução', 84.

21. Sena Sugar, *Moçambique*, passim; Parreira, 'História do açúcar', 262–67 for figures.

22. S. Katzenellenbogen, *South Africa and Southern Mozambique, Labour, Railways and Trade in the Making of a Relationship*, Manchester, 1982, 139–40, passim. Sena Sugar, *Moçambique*, 153–4; Parreira,'História do açúcar', 266.

23. L. Vail and L. White, *Capitalism and Colonialism in Mozambique, a Study of Quelimane District*, London, 1980, 261.

24. Sena Sugar, *Moçambique*, 153–54, 175.

25. Vail and White, *Capitalism*, 257–63; Sena Sugar, *Moçambique*, 84, 151.

26. A. de Almeida Teixeira, *Angola intangivel*, Oporto, 1934, 680; J.M. Cerqueira de Azevedo, *Angola, exemplo de trabalho*, Luanda, 1958, 334–35.

27. T. Heyse, 'Cessions et concessions foncières du Congo', *Congo* (1930), I, 1–7, 17–18; Belgium, Ministère des Colonies, *Plan décennal pour le développement économique et social du Congo Belge*, Brussels, 1949, II, 575.

28. Great Britain, Admiralty, Naval Intelligence Division, *French Equatorial Africa and Cameroun*, London, 1942, 428.

29. J.L. Vellut, 'Originalités et limites de l'industrie manufacturière au Congo Belge, c. 1920–1960', in *Catalogue de l'exposition 'Les Belges à l'étranger: 150 ans de réalisations dans le tiers–monde'*, Brussels,1985.

30. Belgium, *Plan décennal*, 575.

31. Federation of Rhodesia and Nyasaland, *Report of the Commission of Enquiry into the Sugar Industry*, Salisbury, 1962, 2, 4, 54.

32. Vail and White, *Capitalism*, 261.

33. *Companhia do Açúcar de Angola, 4/3/1920 – 4/3/1940*, Lisbon, 1940.

34. Sena Sugar, *Moçambique*, 30–3, 112.

35. E. de Almeida Saldanha, *O Sul do Save*, Lisbon, 1928, xliii.

36. Vail and White, *Capitalism*, 216–21; J.C. Rates, *Angola, Moçambique, S. Tomé*, Lisbon, 1929, 65–71; P. Muralha, *Terras de Africa, S. Tomé e Angola*, Lisbon, c. 1925, 293–98, 307–8. Companhia do Açúcar has interesting photographs.

37. Saldanha, *Mais uma burla*, 20.

38. Vail and White, *Capitalism*, 220, 374.

39. Saldanha, *O Sul do Save*, xliii, for Mozambique in the 1920s.

40. Heyse, 'Cessions et concessions', 17; Companhia do Açúcar, 12–13; G. Lefebvre, *L'Angola, son histoire, son économie*, Liège, 1947, 138.

41. Saldanha, *O Sul do Save*, 191, n. 1, 219; J. F. Head, 'State, Capital and Migrant Labour in Zambézia, Mozambique: A Study of the Labour Force of Sena Sugar Estates Limited', PhD thesis, University of Durham, 1980.

42. Sena Sugar, *Moçambique*, 31–2. Head takes note of this argument, but dismisses it as marginal.

43. Vail and White, *Capitalism*, chapters 6–8. Lefebvre, *L'Angola*, 137–9.

44. D. Abshire & M. Samuels, *Portuguese Africa, a Handbook*, London,1969, 260; Vail & White, *Capitalism*, 384; Vellut, 'Originalités'; Federation of Rhodesia & Nyasaland, *Report*.

18

Employment Practices, Sugar Technology, and Sugar Mill Labour: Crisis and Change in the South African Sugar Industry, 1914 – 1939

David Lincoln

Introduction

For several decades, sugarmill owners in South Africa stood accused by cane–growers of taking an unjustifiably large proportion of the proceeds from sugar sales. The millers and the growers moved very slowly towards a mutual acceptance of principles for realizing (and sharing) surplus value, and it was not until the passage of the Sugar Act in 1936 that stability, if not accord, was reached in the sugar industry over the distribution of sugar revenue. The millers' responses to the related question of producing surplus value were by contrast quite immediate. As they demonstrated during the critical years between the two world wars, employment practices and the pace of technological innovation are acutely sensitive, and susceptible to change, at times of grave realization crises.[1]

Over and above the question of producing surplus value under adverse market conditions, South Africa's sugarmillers had to address the dissolution of the Indian indenture system. It is the responses made by sugarmillers to these two pervasive questions of the period between 1914 and 1939 which form the subject of this essay. In the first part an attempt is made to account for the changing pattern of labour recruitment in the aftermath of the ending of the Indian indenture system. Secondly, technological change and its consequences for sugarmill personnel are considered. In the third part, the sugarmillers' quest for social control is dealt with in conjunction with workers' resistance to exploitation and control.

Labour recruitment

Repeated representations to the government,[2] and a venture to recruit white semi-skilled workers,[3] had by the end of 1917 failed to provide the sugarmillers with an alternative to the Indian indenture system. The sugarmillers were not alone in their predicament, and in January 1918 a joint deputation of sugarmill and Natal mine owners travelled to Cape Town to meet with the Minister of Native Affairs. Consequently the Minister appointed a Departmental Committee to investigate the labour situation in Natal. Sugarmillers were found to have a 40 per cent shortage of African labour, which the Committee associated with the 10 per cent annual decline in the numbers of Indian workers available to them. While supporting in principle the employers' plea to have some assistance from the government, the Committee made a recommendation that was to be oft-repeated to the sugarmillers, namely that by 'sympathetic' treatment the desired response might be elicited from African workers.[4]

By this stage the majority of sugarmillers had joined forces to form the Natal Coast Labour Recruiting Corporation (NCLRC), which recruited African workers in Zululand, Swaziland, and the Transkei, through the medium of white agents.[5] Although some sugarmillers did employ African workers from the British colonies to the north,[6] this was done on a piecemeal basis. There was a strong desire amongst sugarmillers to formalize and extend this practice, and the NCLRC prevailed upon the government to facilitate the recruitment of labour in Mozambique and the tropical British colonies.[7]

One upshot of the sugarmillers' continued lobbying for the government's assistance in ensuring a stable supply of labour was an investigation by the Department of Native Affairs into the conditions facing African workers in the sugar industry. The investigation yielded unprecedented insights into what had been taking place on sugar companies' properties. It was evident that black sugarmill workers were being housed under conditions which reflected the employers' disregard for existing statutes let alone for human dignity.[8] In the same way that insanitary housing had its origins in the employers' efforts to minimize their expenditures on labour, the rations issued to African sugarmill workers reflected economic rather than nutritional priorities.

Government officials waged a relentless battle against employers in the sugar industry over the question of workers' accommodation and rations, and they were given every reason to regard the sugarmillers as recalcitrant reformers. Despite the officials' suggestions as to how some of the more obvious deterrents to potential recruits might have been eliminated, their acrimonious relationship with sugarmillers did little to alter employment conditions. Indeed, the increasing misfortunes of Africans in Natal's rural backwaters gave the profit-minded employers more reason than they had

had since 1913 to ignore injunctions about employment conditions. Such were the depths of those misfortunes that in 1931 the NCLRC went into voluntary liquidation. Starvation in the Reserves was a more efficient recruiting tool than the NCLRC.[9]

Not all sugarmillers were relieved of their labour procurement problems in this way. While the difficulties entailed in recruiting African workers in Natal were subsiding, the anopheles was denying the employers along the north coast and in Zululand the opportunity of exploiting each and every labour recruit to the full. The Umfolozi Sugarmill was severely affected by the ravages of malaria. In 1926 Umfolozi's black workforce had comprised two workers from outside Zululand for every worker from local areas of high malarial endemicity (principally Tongaland, to the north); the sick rate of the former group of workers having been 45 per cent against that of 10 per cent for the latter group. Armed with this evidence, and intent on reducing the time claimed by malaria, Umfolozi had employed only local workers after 1926.[10]

In his 1931 Report on malaria in South Africa, Swellengrebel had cited the Umfolozi statistics in support of his recommendation that the recruiting of workers for the malaria–affected districts be strictly confined to areas of high malarial endemicity.[11] Swellengrebel's recommendation was duly incorporated in the 1934 revision of the Mozambique Convention, which formally allowed employers in Zululand to recruit workers from Mozambique.[12] The Great Depression and the tiny anopheles had accomplished on behalf of the sugarmillers far more than they had succeeded in doing through two decades of purposeful endeavour.

In the early 1900s approximately 90 per cent of all sugarmill workers had been Indians. A little more than a decade after the general Indian strike, Indians had constituted 45.9 per cent of all 7,711 sugarmill employees and Africans 46.3 per cent. By 1935 proportion of Indians had dropped to less than one–third of the now 9,145 employees, while the African share had risen to over 50 per cent.[13] Although this dramatic inversion may have been precipitated by the withdrawal of Indian workers, it had been accomplished against a background of spatially uneven development and more active involvement by sugarmillers in the regional labour market. On the one hand, the ability of peasant and petty commodity production to support an expanding reserve populace had been diminishing; on the other hand, sugarmillers had succeeded in replacing the indenture system with oscillating labour migrancy. Organized and aggressive recruitment had carried the sugarmillers through the critical twenties, but not without intense struggle. However, the coalescence in the early thirties of ecological disaster and general economic crisis had stripped local Africans of the means of avoiding the distressing conditions of sugarmill work. The sugarmillers, who were thus relieved of pressure in the recruiting stakes, could be expected to remain complacent about the consequences of keeping down their expenditures on variable capital.

Technical change

After war–time restrictions on sugar sales were lifted, most sugarmillers paid close attention to technological change.[14] The more important changes were made in the mill house, with the objective of improving performances in the extraction stage of the production process. Largely because of the ubiquitous cultivation of Uba cane in Natal,[15] special grooving was required in mill rollers, and milling tandems usually incorporated at least one single or double crusher before the mills themselves. Milling tandems now commonly had one or two sets of knives installed between the crushers and the mills. In some of the largest sugarmills this combination of knives, crushers, shredders, and milling train was duplicated so as to have two tandems operating simultaneously; the advantages thereof being reduced breakdown or overhaul time, as well as greater throughput and higher extraction rates. The responsibility for maintaining and controlling this range of mechanical devices, and attendant workers, was shouldered by the engineers.

The role of the sugarmills' engineers as custodians of mechanical plant and its operatives was still in the mid–1920s defined on the basis of labour–using rather than labour–saving principles. Despite the tempo of new investment in mechanical plant, manual labour was advocated as an important element of the production process. The view was still held in some quarters, for example, that, 'the old, if expensive method of hand feeding [cane into the mills] is undoubtedly the best. By hand feeding may be secured that even supply of cane which makes for regular crushing'.[16] As far as the training of workers was concerned, neither indentured nor migrant African workers were considered worthy of expenditure. Thus it is perhaps not surprising that in 1926 the Board of Trade and Industries should have detected a generally 'inefficient' use of labour by sugarmillers, and related it to the manner in which 'the executive minds of the industry have not entirely thrown off a certain bias in favour of low–paid labour and a consequent tolerance of its shortcomings'.[17]

What the Board's investigation implicitly revealed was that skills and relative costs determined the way in which workers of different 'racial' categories were assigned places in the technical division of sugarmill labour. Whites were seen in general to have replaced 'coloureds' (presumably Mauritian settlers) in the boiling house and engineering departments. On the other hand the Board found that Indian centrifugal attendants had been replaced by Africans rather than by whites; something which was considered inappropriate, except in Zululand where the climate might 'render this type of work somewhat trying for Europeans'.[18] White workers were generally advantaged, although the sugarmillers' quest for cheaper labour did not allow for formal colour bars particularly with regard to semi– and unskilled workers.

Insofar as the Board of Trade and Industries' inquiry led to the Fahey Conference of millers and growers, its report presaged a revision of employers' practices in the sugar house, where cane juice is processed. The application of analytical chemistry to control sugar processing had been neither universal nor sophisticated in the South African sugar industry before the crisis of the mid–1920s.[19] The 1926 Fahey Conference provided the catalyst for realizing the potential of chemical control. Amongst other measures adopted, with the objective of introducing a more equitable system of sharing sugar revenue between millers and growers, the Fahey Conference Agreement embodied a commitment to establish a testing service to monitor growers' cane from the time that it was delivered in the mill yard.

The basis for the growers' cane–testing service was laid by the Moberly Report of 1927, which suggested a hierarchy of personnel comprising supervising chemists earning £25 per month, testers earning a minimum of £12 per month, and 'young reliable Indians' as laboratory assistants earning £4 per month plus rations.[20] This implied that a training programme had to be instituted in order to prepare some 50 growers' chemists for their task of vigilant analysis at the sugarmills. Because the sugarmillers had recognized the importance of chemical control they were interested in the establishment of common training facilities for growers' chemists and their own sugar house apprentices. However the sugarmillers were remarkably hesitant about making any expenditures towards the scheme, and when the four–year course was inaugurated in 1928 it was left to lecturers paid by the South African Cane Growers' Association to provide theoretical training for cane–testers as well as sugar house personnel during the annual off–crop.[21]

While the new imperative of chemical control meant an increase in the number of laboratory and sugar house personnel under the supervision of chemists and factory overseers respectively, there were also now possibilities for reducing the number of workers employed in the mill house. Control over prime movers could now be taken from the hands of individual drivers to become increasingly centralized and even electrically operated.[22] Such innovations were not necessarily inspired by labour–saving ideals, but they were heralded by owners and sugar technologists in terms which indicated that they were poised to review their thinking on the utilization of labour. Although in the late 1920s the South African Sugar Technologists' Association had established a committee to investigate labour–saving, the spur towards a systematic policy of labour–saving in the sugarmills was provided by anticipated legislation for an eight–hour day. Such legislation, it was noted by the Association's president in his 1937 address, would make it necessary to reduce labour costs.[23]

The pace of technical change during the interwar period had followed business cycles very closely. In the first place, the promise of high returns on sugar sales in the early years, before 1925, had created the incentive for increasing the mills' crushing and throughput capacities. Secondly, the crisis of the mid–1920s had given sugar producers good cause to enhance realization circumstances. The rapid implementation of chemical control and the subsequent focus on labour–saving had taken place in the wake of that crisis and within the stimulating context of repeated instances of low–priced foreign sugars being dumped on the domestic market.

Social control and resistance

In 1914 there began the process whereby the 'racial' division of sugarmill labour would ultimately be partly inverted; that is, in so far as African workers replaced Indian workers as the numerically preponderant group amongst sugarmill employees. It was a partial inversion in that the cleavage between black and white workers was maintained. This fundamental characteristic of the 'racial' division of sugarmill labour was reinforced by the employers' policies and practices vis–à–vis black and white workers respectively, notably in the sphere of control. As might be expected, workers' responses to differential treatment followed separate trajectories; white workers leaning towards a conciliatory and acquiescent mode of relating to the sugarmillers, and black workers being predisposed to taking up the cudgels against their employers in various individual or collective acts of resistance.

The first recorded collective act of resistance by sugarmill workers during the period occurred at the Umfolozi Sugarmill on 25th September 1918. At 7 a.m. on that day all 260 African workers downed tools, demanding that their monthly wages be increased from £2 to £3. When, about an hour later, their demand was reportedly raised to £4, the manager informed the striking workers that he was calling in the police. The threat of police action was itself persuasive, and the strike was called off several hours before the arrival at 5 p.m. of a police contingent from Empangeni. The Umfolozi strike was ill–fated, and within a month over 100 of the frustrated workers whose demands had not been met deserted the sugarmill.[24] It was evident that their employers had had no intention of entering into any form of negotiation with them. It was equally apparent that the workers had seen retreat and withdrawal as the most feasible option to exercise within the constraints of their own vulnerability.

The remoteness of Umfolozi's location in Zululand was illustrated by the time which elapsed between the dispatch of the manager's telegram to the magistrate in Empangeni and the arrival of the policemen at the sugarmill. Not every sugarmill was quite so remote, but there were very few which were not isolated from one another and from urban settlements by considerable distances. Such isolation bedeviled attempts to organize sugarmill workers, and it made them vulnerable to employers' machinations.

White workers were the first to be paternalistically embraced by their employers in a campaign which was launched during the prosperous period (for sugarmillers) between the end of World War I and the 1925 sugar crisis. Its implicit purpose was to inculcate a sense of community, and thus allegiance, amongst white sugarmill personnel. Community–building involved the sugarmillers in a wide range of tactics covering all aspects of life in the sugar villages. There were measures taken to solidify the white employees' community by granting them share options and bonding them in an economic sense with their employers.[25] Religious practice was encouraged through the provision of a site or funds for the construction of a church.[26] By also building recreation halls in the sugar villages the sugarmillers established a balance between the sacred and the profane amenities available to their white employees.[27] The owners of the Tongaat Sugar Co. may well have been led to believe that the community–building exercise bore immediate benefits when they found cause to praise 'all members of the trade unions employed by the Company (with the exception of 6 apprentices and labourers) [who] decided to resign from their respective unions rather than resign from the Company with its privileges to become daily paid labourers at the union rates of pay'.[28]

The opening of a new recreation hall usually called for the presence of some company figurehead. It is unlikely that any of these ceremonies was more auspicious than when Smuts opened the Zululand Sugar Milling Co.'s hall while he was on a tour of Zululand in 1922;[29] all the more so because the blood had barely dried since Smuts's troops had crushed the recent white mine workers' strike.

By 1924 the sugarmillers had progressed some distance along the community–building trail. But whatever success they had achieved in nurturing the allegiance of their white employees, the sugarmillers had seen no reason to extend their campaign to embrace black workers. It would have been quite consistent with this attitude towards black workers if the sugarmiller J.L.Hulett had failed to attach due significance to the first meeting to be convened in Durban by the Industrial and Commercial Workers' Union of Africa (ICU). The irony of the matter was that it was in Hulett's Durban home that Clements Kadalie was granted official permission in 1924 to hold the meeting.[30]

Following its Durban debut, the ICU gained widespread support amongst Africans in Natal. While it is clear that some of this support came from along the coast,[31] the sugar belt never became an ICU stronghold.[32] Nevertheless the sugarmillers did not take kindly to the

ICU's presence in the sugar belt, and they were anxious enough about the Union's advances to deem obstructive tactics necessary.[33] Both Helen Bradford and Shula Marks have shed light on the sugarmiller W.Campbell's connivance in mounting an indirect attack on the ICU in 1927 through the agency of his acquaintances Solomon, the royal Zulu paramount, and John Dube. After Dube had spoken at Campbell's invitation, and in Solomon's presence, to a thousand workers at Mount Edgecombe, he augmented his verbal condemnation of the ICU and its Natal secretary in his newspaper *Ilanga lase Natal*.[34]

The sugarmillers were astute politicians with a keen sense of the role which Zulu royalty, ideologues like Dube, and figureheads of the ruling classes like Smuts, could be called upon to play in making direct as well as subtle appeals to workers' sympathies. They recognized the symbolic importance furthermore of comporting themselves as paternal employers. While the 'father'/employer had continued during and after the Great Depression to heap actual or symbolic favours upon white employees, the first stirrings could now be felt of a movement which was to stimulate organization amongst black sugarmill workers. Since 1935, but especially after the 1937 dispute at the Durban Falkirk Iron Co.'s foundry, the organization of black industrial workers in Natal was spearheaded by members of the Communist Party of South Africa (CPSA).[35] This did not augur well for the sugarmillers, for in 1937 the Natal Sugar Industry Employees' Union (NSIEU) was established under the leadership of legendary organizers such as H.A.Naidoo and G.Ponnen, two of the earliest Indian members of the CPSA.[36]

The NSIEU was constituted and registered as the first union of Indian sugarmill and refinery workers, with a 'parallel' but officially unrecognized African section. The birth of the Union, at the very time that labour-saving had come to be close to the forefront of the sugarmillers' thinking, was a significant event. However its influence over sugarmill workers was limited on two counts. The NSIEU did not have a large base of members in the sugarmills, and it did not offer an immediate means of eliminating the effects of the general economic situation and related pressures upon sugarmill workers at the point of production.

Labour-saving meant different things to different people in the sugar industry. When two of Doornkop Sugarmill's African workers stood trial in mid–1937 on charges of malicious injury to property (by setting fire to company cane), the court found *inter alia*:

> Dissatisfaction among employees...as a result of alteration of conditions of employment. ...Ill feeling on part of malcontents against those employees who had accepted the new terms.... Rumour current amongst employees that those refusing the new terms would be dismissed....On 20/5/37 twenty five acres of cane

on the estates were burned.... As a result of the fire the Mill started crushing immediately and, as all employees were needed for the extra work thus entailed, none of the malcontents were dismissed.[37]

One of the accused was acquitted, the other convicted. The latter worker, an *induna* [African foreman] in the sugarmill, had refused to accept the lengthening of the working month by his employers.[38] His ingenious act of resistance was representative of the spontaneous workers' responses which were to persist in the sugarmills, especially where the NSIEU had no presence.

Some two years after the conviction of the Doornkop *induna*, nine of Felixton Sugarmill's African workers, including three from Mozambique, were charged and convicted for inciting 76 of their African colleagues to withhold their labour. The charges arose out of the refusal by an entire shift of African workers to work one night early in June 1939 unless their monthly wage of £2 was increased. Their current wage was the same as that which Umfolozi sugarmill's African workers had regarded as inadequate, two decades earlier. Unlike Umfolozi's workers, and disregarding appeals made by the factory manager and the compound manager in the presence of a police contingent, the striking workers at Felixton stood fast and the sugarmill had to be closed down for the night. As in the case of the Doornkop 'malcontents' whose resistance landed them in court, Felixton's striking workers were the losers in their efforts to contest the *status quo* in the sugarmill.[39] With the list of such defeats growing, the limited reach of the NSIEU continued to be shown up particularly where sugarmills furthest from its Durban headquarters were concerned.

An important factor which accounted for the NSIEU's limited reach and organizational strength, was that from the outset its work was hampered by employers. Union meetings were prohibited on company property, and clandestine meetings were under constant threat of disruption by armed intimidators.[40] Consequently, during the early years of the NSIEU's existence its officials travelled long distances to remain in touch with workers while obviating the victimization that was expected if local stewards were appointed.[41]

If their practices towards sugarmill workers had gone no further than stark economism, and had been based on expectations of unquestioning submissiveness, the sugarmillers had nevertheless approached white workers and black workers in manifestly different ways. The treatment received by white workers, who had been the subjects of their employers' community–building campaign, had betokened their status in the technical division of labour as much as their 'racial' status. Community–building had been instituted as a means of social control at a time when sugar technology was rapidly advancing and new positions were opening up for supervisory personnel and technologists. Moreover, the growth in the numbers of whites recruited by the sugarmillers, and the privileges they were accorded, had been consistent with the 'civilized labour' policies which had evolved in a climate of white impoverishment and heightened militancy amongst white mine workers. 'Uncivilized' or black, sugarmill workers had not been favoured with the same tactical reactions by their employers to social change and struggle at the national level. Indeed, the sugarmillers had had the political wherewithal to displace the burden imposed by the realization crisis onto the backs of the black majority of the workforce. Consequently, by the late 1930s (when black urban workers were widely attempting to stave off the ravages of inflation by means of organized militancy) black sugarmill workers had begun to arm themselves with the weaponry of organization. Both working conditions and living conditions presented a multitude of issues which could reasonably have been expected to stimulate the black workers' interest in organization, but the NSIEU would have to bide its time before its constituency was solid enough to justify a frontal assault to seek redress on these matters.

Conclusion

For South African sugar capital, the period between the outbreak of the two World Wars was a protracted test of resilience and resourcefulness. Repeated realization crises of global proportions, compounded by the local labour crisis, necessitated extraordinary action on the part of sugarmillers. Their responses which were brought to bear on the sugarmills and on conditions in their attendant villages ultimately revolved around two concerns; namely, to secure permanent access to a means of satisfying the need for menial and unskilled workers, and to raise the productivity of sugarmill labour through technological innovation and the active promotion of stable social relations. Although they were generally successful, their endeavours began to be complicated towards the end of the period by incipient organization amongst black sugarmill workers. But, by the outbreak of WWII this organization had not become significant enough to pose a real threat to the prospects for capital accumulation.

References

41. In making explicit this analytically useful distinction between the production and the realization of surplus value, I have taken the cue from

C.R. Littler and G. Salaman, 'Bravermania and beyond: recent theories of the labour process', *Sociology*, 16 (1982), 251–269: 'Surplus value has to be produced but also realized in the market. What this implies is that the realization of surplus value (i.e. finding markets, selling in those markets and making a profit) may be more crucial than the production of surplus value for certain firms, certain industries or during certain periods.'

2. For example, Transvaal Archives Depot (TAD), GNLB 81, Resolutions passed by Natal Sugar Assoc. Labour Committee, 1913; *Ibid.*, Notes of meeting between Dir. of Native Labour and sugar planters, 15 Dec. 1913; and *Ibid.*, Notes of meeting between Prime Minister and Natal Labour Assoc., 16 March 1914.

3. TAD, MNW 238, Saunders to Smuts, 24 Dec. 1913; *Ibid.*, Saunders to Sec. of Mines and Industries, 7 Dec. 1914; *Ibid.*, Smith to employers, 12 Feb. 1915; *Ibid.*, Campbell to Smith, 15 Feb. 1915; *Ibid.*, Armstrong to Smith, 19 Feb. 1915; and *Ibid.*, Assist. Inspector White Labour to Sec. for Mines and Industries, 3 May 1915.

4. House of Assembly (Union of South Africa), Annexure 53, 1919.

5. TAD, GNLB 252, Shortage of Native Labour Committee's Durban evidence, 22 June 1918.

6. Natal Archives Depot (NAD), CNC 1916/1874, Passes inward, 1916; *Ibid.*, CNC 1917/2105, Natal Estates' application, 11 April 1917; and TAD, GNLB 308, Notes of meeting, 26 Nov. 1919.

7. TAD, GNLB 252, NCLRC to Sec. for Native Labour, 23 Dec. 1918 and 3 Jan. 1919.

8. TAD, GNLB 308, Dundee Native Affairs Dept. Inspector to Dir. of Native Labour, 4 Dec. 1919.

9. It was of particular concern to the Dept. of Native Affairs that the depression and drought had raised unemployment levels on the Witwatersrand, and Native Commissioners were instructed to restrict the flow of Africans to the mines. NAD., Lower Umfolozi Magisterial Archives, 3/4/9/1, Sec. for Native Affairs' circular, 25 Jan. 1932; *Ibid.*, Sec. for Native Affairs' circular, 11 Feb. 1932; and *Ibid.*, Dir. of Native Labour's circular, 17 March 1933. Also *South African Sugar Journal (SASJ)*, 15 (1931), 695; and NAD, Lower Umfolozi Magisterial Archives, 3/4/9/1, Chief Native Commissioner's circular, 11 Dec. 1931.

10. Union of South Africa, *Report on Investigation into Malaria in the Union of South Africa 1930–31*, Pretoria, 1931.

11. *Ibid.*

12. S.T. van der Horst, *Native Labour in South Africa*, London, 1942, 288; and for official interpretations of what the revision implied for employers in Zululand, NAD, Lower Umfolozi Magisterial Archives, 3/4/9/2, Chief Native Commissioner's circular, 1 March 1935. The revision also

created a conduit for workers from Mozambique who went to the Witwatersrand via the sugar belt. NAD, Lower Umfolozi Magisterial Archives, 3/4/9/3, Sec. for Native Affairs' circular, 27 Nov. 1935; *Ibid.*, Chief Native Commissioner's circular, 30 April 1936; and *Ibid.*, Chief Native Commissioner's circular, 19 May 1936.

13. Colony of Natal, *Statistical Year Books*; Board of Trade and Industries, Report no. 66, *Report on the Sugar Industry,* 1926; *SASJ*, 19 (1935), 201-03.

14. R.F. Osborn, 'Sweet Concord', unpublished ms., nd, South African Sugar Assoc. Library.

15. G.S. Moberly, 'The replacement of Uba by new variety canes from 1936 to 1944', *South African Sugar Technologists' Association Proceedings,(SASTA),* 19 (1945), 29–34; and M. McMartin, 'The early days of the Natal sugar industry, with special reference to the introduction of varieties', *SASTA,* 22, (1948), 83–9.

16. R.F. Hutcheson, 'The milling of Uba cane in Natal', SASJ, 10, (1926), 165-71.

17. Board of Trade and Industries, Report no. 66, *Report on the Sugar Industry,* 1926.

18. *Ibid.*

19. Osborn, 'Sweet Concord', 364–65.

20. *SASJ*, 11 (1927), 147–49.

21. A. Hammond *et al., South African Cane Growers' Association: The First 50 Years,* Durban, 1977, 82.

22. For example *Natal Mercury,* 15 June 1932; and J.B.M. Godfrey, 'Electrical Activities of Natal Estates', *SASTA,* 8 (1934), 103–28.

23. 'President's Address', *SASTA,* 12 (1937).

24. NAD, CNC 1918/33, Rumbelow to O.C.C. Eshowe, n.d.; *Ibid.*, Rumbelow to Empangeni Magistrate, 2 Oct. 1918; and *Ibid.*, Empangeni Magistrate to Chief Native Commissioner, 24 Oct. 1918.

25. *Condenser*, 5 (1967), 3.

26. For example, *SASJ*, 4 (1920), 461; *Ibid.*, 17 (1933), 72.

27. The provision of a recreation hall had earlier been perceived by mine owners on the Witwatersrand as a means of promoting social control amongst white workers. C.van Onselen, *Studies in the Social and Economic History of the Witwatersrand 1886–1914*, Johannesburg, 1982, I, 39–40. There is every likelihood that this served as an example to the sugarmillers.

28. *Condenser*, 5 (1967), 3.

29. *SASJ*, 6 (1922), 659.

30. C. Kadalie, *My Life and the I.C.U.*, London, 1970, 64.

31. As early as 1925 the ICU had convened a meeting 'at Zululand' (Forman Papers, University of Cape Town Library, B.C. 581, B1.5, Notice of I.C.U. Executive Committee, C. 6 Oct. 1925); and by 1929 there was certainly a branch of the ICU in Stanger (*Ibid.*, B5.35, Champion to Mdima, 23 July 1929).

32. H. Bradford, 'The Industrial and Commercial Workers' Union in the Natal Countryside: Class Struggle on the Land', unpublished Hons. dissertation, University of Cape Town, 1980, 50.

33. *Ibid.*, 148, 372.

34. H. Bradford, 'The Industrial and Commercial Workers' Union of Africa in the South African Countryside', unpublished PhD thesis, University of the Witwatersrand, 1985, 148; *Ilanga lase Natal*, 12 Aug. 1927.

35. E. Roux, *Time Longer than Rope: The Black Man's Struggle for Freedom in South Africa*, Madison, 1964, 250–52, 330–31.

36. K. Luckhardt and B. Wall, *Organize or Starve! The History of the South African Congress of Trade Unions*, London, 1980, 78.

37. NAD, Lower Tugela Magisterial Archives, Criminal Case 1431 of 1937.

38. *Ibid.*

39. NAD, Lower Umfolozi Magisterial Archives, Criminal Case 568 of 1939.

40. Luckhardt and Wall, *Organise or Starve*, 78.

41. H.G. Ringrose, *Trade Unions in Natal*, Cape Town, 1951, 73.

INDEX